ION ASSISTED SURFACE TREATMENTS, TECHNIQUES AND PROCESSES

WITHDRAWN

Proceedings of the
conference sponsored and organized by
The Metals Technology Committee of The Metals Society,
co-sponsored by The Welding Institute and the
Surfacing Division of The Welding Institute,
and held at the University of Warwick,
Coventry, 14–16 September 1982

THE METALS SOCIETY
1982

Book 288
published by
The Metals Society
1 Carlton House Terrace
London SW1Y 5DB

©1982 The Metals Society

ISBN 0 904357 48 1

Prepared from authors' typescripts and original illustrations

Printed and made in England by The Chameleon Press Ltd, London

DJ&TGH   SECOND IMPRESSION SPRING 1984

CONTENTS

T TAKAGI

Ion plating and ion beam deposition

SYNOPSIS

In ion plating and ion beam deposition, film properties can be controlled by changing the deposition conditions such as the acceleration voltage and the content of ions. Fundamental effects of these parameters in different ion-assisted deposition methods were discussed. In the ion plating, kind and type of the impinging particles on the substrate surface are much more complicated than in the ion beam deposition. A border between the ion plating and the ion beam deposition is determined by the gas pressure of the working region and is in the range of 10^{-4} - 10^{-3} torr. The effective energy of the incident particles in various kinds of film formation techniques is of the order of a few eV to a few hundreds eV. A profound effect caused by the presence of the ion was confirmed even though the amount of charged particles is less than a few percent in total flux. They affect sputtering, sticking coefficient, adatom migration, nucleation, chemical reaction, etc. Several examples were reviewed to characterise the deposition features.

THE AUTHOR

The author is with Department of Electronics and Ion Beam Engineering Experimental Laboratory, Kyoto University, Sakyo, Kyoto, Japan.

1. Introduction

Ion plating and ion beam deposition offer control over fundamental characteristics of film growth not possible by conventional film formation processes. That is, film growth conditions can be controlled by changing deposition parameters such as the type and kinetic energy of impinging ions.

The fundamental effects of kinetic energy of the ions on film formation are sputtering, implantation, formation of nucleation sites, heating, enhancement of surface atom migration, etc. The presence of the ionized particles in the evaporant beam, even without acceleration of the ionized particles or even when only a few percent of ionized particles are included, greatly influences the critical parameters of the condensation process for film formation and enhances chemical reaction.

Generally, surface interaction of ions with a substrate are extremely complex so that a full analysis of fundamental effects during growth has not been done. This is especially true in deposition by ion plating where ions, high energy neutrals and electrons are involved in the deposition process. Therefore, besides the effects due to

high energy neutral atoms, their radicals must be taken into account. In this method, ions produced in a plasma will experience many collisions before arriving at the cathode and lose their energy in charge transfer collisions whereby energy is transfered to neutral atoms producing a lot of energetic and excited neutral species.

In ion beam deposition, film growth takes place in a relatively higher vacuum region than that in ion plating. The kind, the energy and charge state of particle impinging on the substrate surface are more controllable, leading to better film deposition conditions. The border dividing the two methods may be described by the gas pressure in the working region and is in the range of 10^{-4} - 10^{-3} torr where mean free path of the gas in the chamber is comparable with the distance between the ion source and the substrate. Differences in deposition processes and their characteristics will be mainly due to the above conditions.

In this paper, we summarize and review various kinds of experiments on film formation to analyze the fundamental role of ions during the process. To understand further details of the role of ions, fundamental deposition parameters on film growth are studied by ionized cluster beam (ICB) deposition in connection with the kinetic energy and ionic charge. Lastly, characteristics of films formed by ionized cluster beam will be described.

2. Role of ions in film formation

In an actual deposition, film properties are considered to be influenced by the following conditions; 1) the arriving rate and kinetic energy of the deposition vapour, 2) the substrate temperature, 3) the ambient gas pressure, 4) surface contamination, 5) the presence of electric charge on the vapour and/or the substrate, 6) structural defects on the surface, 7) the resulting creation of surface defects. In an ion-assisted film formation, ions transfer energy and charge to a substrate and a depositing film surface. The role of ions becomes primarily important and may be described in terms of kinetic energy and ionic charge.

2.1 Kinetic energy

Fundamental effects of the kinetic energy of ions on the film formation[1,2] are listed in Table 1. These effects influence the crystallographic and physical properties of the films. In the ion-assisted deposition, adhesion strength, packing density, surface roughness, crystalline state and structure of the deposited films, etc. may be affected considerably by the acceleration of ionized particles.

An upper limit of the acceleration voltage may be determined by the sputtering yield. An

Table 1 Influence of kinetic energy on film formation

Fundamental effects of kinetic energy	Influence on film formation
Surface cleaning by sputtering	Improvement of adhesion Removal of the surface oxide or contaminated layers just before the deposition
Deep etching	Mechanical improvement of adhesion
Blending of sputtered material with incident evaporant particles	Formation of interfacial layer
Creation of suitable amount of activated centers such as defects and displacements of surface atoms which act as centers of nucleus formation	Enhancement of the growth of nuclei at the initial stage of film formation Formation of an interfacial layer and increase in bonding energy between substrate and deposited atoms
Suitable ion bombardment and sputtering during deposition	Change in the morphology Stimulation of nucleation, growth of nuclei and coalescence
Ion implantation (including recoil implantation)	Enhancement of interfacial layer formation
Heating by the thermal energy converted from kinetic energy	Change in the morphology Increase in the chemical reactivity
Migration of depositing particles on the substrate	Enhancement of surface diffusion energy keeping a relatively low substrate temperature, resulting in change in morphology or growth of epitaxial film

Table 2 Optimum conditions for the kinetic energy of ions incident on the substrate for film formation

Condition	Required incident ion energy	Results
Deposition	Less than the energy corresponding to the sputtering rate $(S(E)=1)$ Larger than the energy at which the sticking probability becomes too low	
Surface cleaning	Larger than the energy of adsorption on the substrate surface, i.e. 0.1–0.5 for physically adsorbed gases and 1–10 eV for chemically adsorbed gases	Optimum value of kinetic energy: a few to a few hundreds eV
Good quality film formation	In a range where enhanced adatom migration influences properties of the deposited film suitable ion bombardment affects the growth of nuclei a suitable amount of defects or atomic displacement near the substrate surface contributes to film formation during the initial stage	

empirical formula of the sputtering yield by the low energy ion beam near the threshold is given as follows [3,4]:

$$S(E) = (0.42\alpha/U_s)K \cdot S_n\{1-(E_{th}/E)^{\frac{1}{2}}\}$$

where α is a factor depending mainly on the ratio of the mass of the target atoms M_2 and the mass of the incident ion M_1, U_s is the heat of sublimation (eV), S_n is the nuclear stopping power in the reduced units and E_{th} is a parameter related to the threshold energy (eV). K is a constant given by

$$K = 8.48 \frac{Z_1 \cdot Z_2}{(Z_1^{2/3} + Z_2^{2/3})^{\frac{1}{2}}} \cdot \frac{M_1}{M_1 + M_2} (\frac{eV \cdot cm^2}{10^{15} \ atoms}).$$

Several examples are calculated and are shown in Fig. 1. The beam energy necessary for the deposition must be sufficiently below the condition of S(E)=1.

Too much decrease in the kinetic energy of the ion generally leads to a decrease in the sticking coefficient. To determine the lowest energy for the deposition, threshold energy for defect production on the substrate surface can be applied as one of ideal conditions.

Since a probability that the target atom obtains energy between y and y + dy by a hard sphere collision is given by $1/T_m \cdot dy$, the probability to transfer energy higher than E_d to the target atom is expressed by

$$\int_{E_d}^{T_m} \frac{1}{T_m} \ dy = \frac{T_m - E_d}{T_m}$$

where T_m is the maximum energy transferred to the target atom[5], E_d is a displacement energy of the target atom. Cross-section of the displacement by the hard sphere collision is expressed by

$$\alpha = \pi R^2 (1 - \frac{E_d}{T_m}) = \pi R^2\{1 - \frac{(M_1 + M_2)^2}{4M_1 M_2} \cdot \frac{E_d}{E}\}$$

where E is incident energy of ion. R is the effective radius of colliding atoms and is given by the solution of the following equation:

$$E = \frac{Z_1 Z_2 e^2}{R} \exp(-R/a)$$

where Z_1 and Z_2 are atomic number of the incident ion and target atom, respectively, and a is the screening distance.

As an ideal deposition condition, the energy of ion beam can be chosen higher than the thermal energy of the order of a few tens meV and lower than that of defect production. Figure 2 shows the calculated cross-section of displacement for Si target by Ag and Si ions. Energy lower than E_d which gives defect-free deposition condition is still high enough compared to the thermal energy.

If the energy of ions is larger than the displacement energy of the substrate atoms, the substrate will suffer a series of atomic collisions and defects which act as nucleation sites will be formed. Desorption of adsorbed atoms or molecules are also expected. Physical adsorption energy of atoms or molecules is usually smaller than the energy of chemisorption where electron transfer is involved. Since binding force of the inactive atoms or molecules is the van der Waalse force, the energy is commonly in a range less than a few tenth eV. These adsorbed atoms can be easily desorbed even by the bombardment with a low energy ion beam.

Enhanced surface mobility is another important factor on growing process. The effect is caused by the momentum-transfer-induced diffusion of impinging materials upon bombarding the substrate surface, resulting in increase in adatom migration.

The factors to determine the optimum conditions are listed in Table 2. Taking account of all considerations, an optimum range of the kinetic energy of the particles is from a few to a few hundreds eV.

2.2 Ionic charge

The charge derived from ion beam and/or electric field at substrate interface affects the film formation condition. Electrically charged island increases the free energy and the increased energy will be accommodated by an increase in the surface area, which gives rise to coalescence at an earlier stage of the film formation. The electrostatic force between charged islands also enhances the migration and coalescence of the islands[6].

Effects caused by the ionic charge of the impinging ions without acceleration voltage are seen in elemental and compound film formation such as Ag, Au, Cu, ZnTe on NaCl[7,8], even though the content of ion is less than a few percent in total flux.

Another effect of ionic charge is the enhancement of chemical reaction at the substrate surface. Synthesis of compounds such as oxides, nitrides and carbides was found to be possible at relatively low temperature in reactive ion plating and ion beam deposition. Preferred oriented ZnO[9] and BeO [10] films are formed to 200 - 400 °C by reactive ionized cluster beam (RICB) deposition without applying acceleration voltage.

Main effects of presence of ionic charge are summarized in Table 3.

In actual film deposition, there are so many interdependences among the fundamental parameters controlled by the deposition conditions. Especially when high energy neutrals, metastable atoms and radicals are concerned, much attention must be given to analyze the deposition process.

3. Characteristics of various kinds of ion plating and ion beam deposition

Since ion plating method is essentially a combination of the evaporation and the use of a glow discharge, developments have been made in the design of various types of evaporation source including sputtering sources and of the ionization system. They are reviewed in recent paper[11]. Resistance heating, flash evaporation, electron-beam heating, hollow cathode evaporation, r.f. induction heating and sputtering are used as the evaporation sources. The use of different kinds of evaporation method has not found much difference in the fundamental characteristics in deposition mechanism unless the vapour pressure of evaporant is not enough to affect discharge. On the other hand, ionization efficiency of ionization system has strong influence on the quality of the deposited films owing to the deposition conditions such as the percentage of ions in the total flux, bias voltage, inert or reactive gas pressure, etc. A use of r.f. coil as an ionizer extends operation region to a high vacuum region of the order of 10^{-4} torr[12]. A triode ion plating system also increases an ionization efficiency and can lower the possible operation pressure[13]. In ion plating, the substrate surface is subjected to a flux of energetic ions and energetic neutrals. Lowering the gas pressure during the deposition reduces content of energetic neutrals and increases intensity of ion beam. The high vacuum operation condition improves

Table 3 Main effects of ionic charge for film formation

Main effects of ionic charge	Influence on film formation
Effects on critical parameters at the early stage of film formation	Stimulation and enhancement of nucleus formation, nucleus growth and coalescence Reduction of critical condensation pressure for film formation Alterations of capture cross-section of adatom Enhancement of preferred orientation and reduction of substrate temperature necessary for crystalline film growth Increase of adhesion strength and formation of a uniform film (increase of substrate-film binding energy and decrease of internal stress of a film)
Enhancement of chemical activity	Formation of oxide, nitride or carbide films by reactive deposition Activation of compound synthesis

These effects are remarkable even if no additional acceleration voltage is applied or number of ions is small.

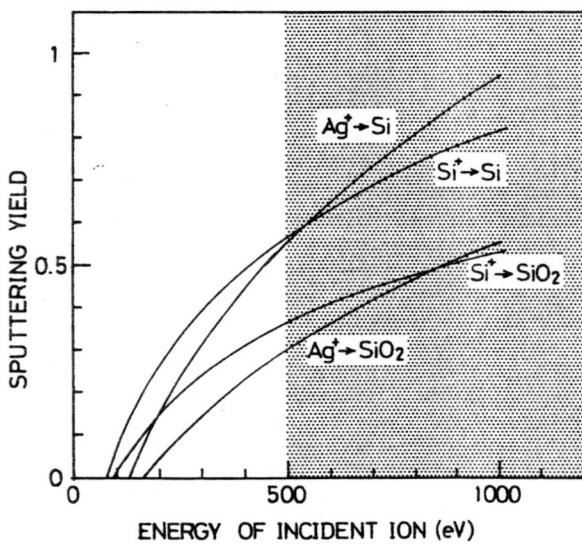

Fig.1 Sputtering yield by low energy ion beams

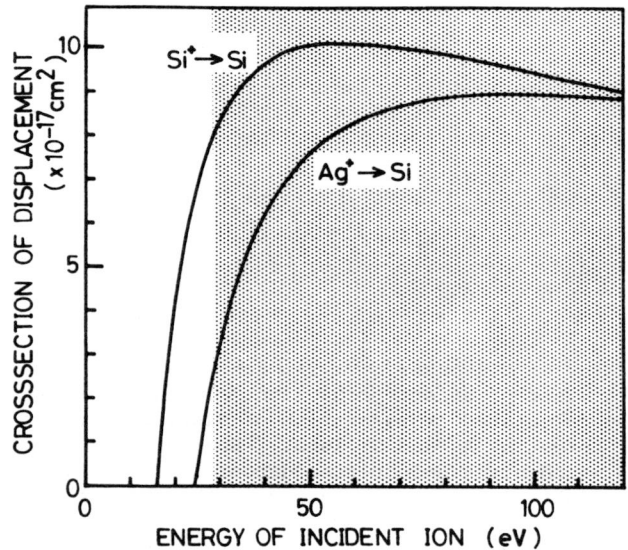

Fig.2 Cross-section of displacement by low energy ion beams

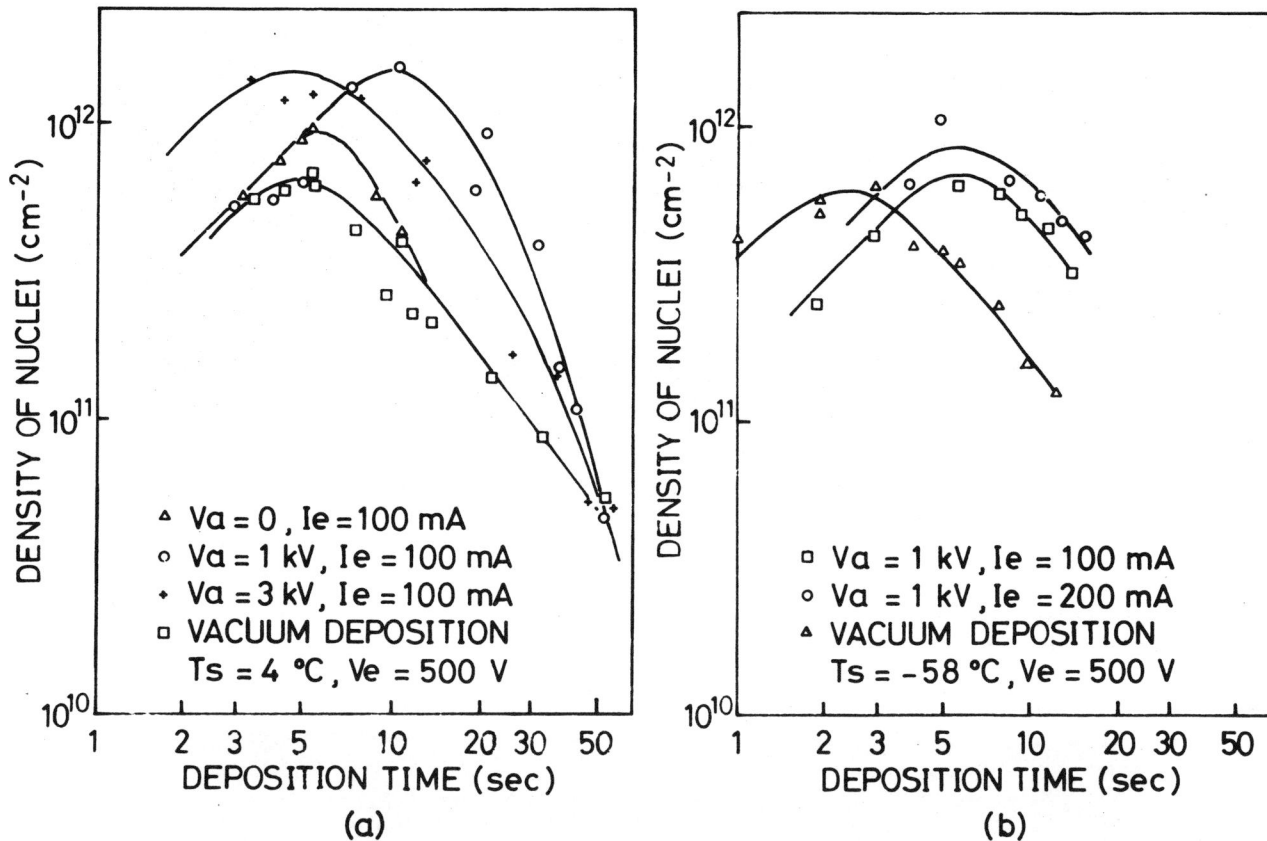

Fig.3 Number of nuclei as a function of deposition
time

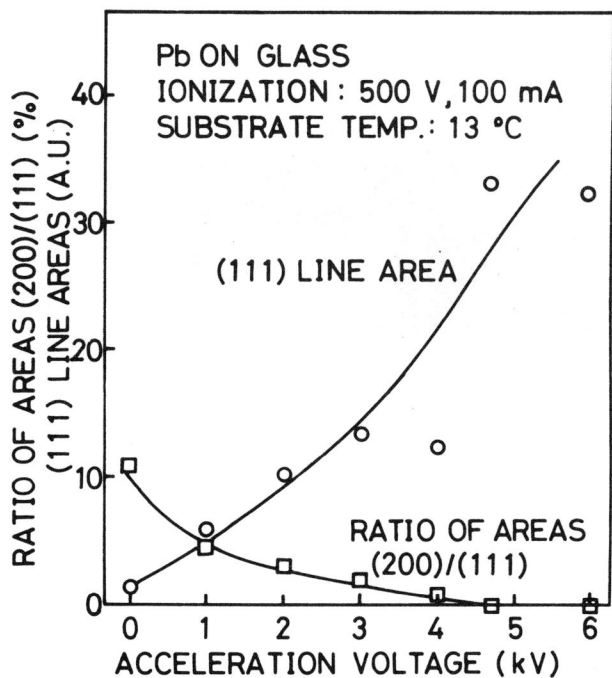

Fig.4 Dependence of (111) orientation on
acceleration voltages.

film characteristics considerably. Dependence of
adhesion and crystallinity on plasma potential in
magnetron sputtering type ion plating with additional electron emitter have been studied[14]. The
results show that adhesion strength and crystallinity increase proportionally to the plasma potential in the experimental conditions. Details of
the film formation mechanism in ion plating process
are not fully understood; however, it can be estimated that film is affected by a small number of
energetic ions and a large number of energetic neutrals, even in high vacuum operation conditions.

Simple estimation of the type and the energy
in ion plating process has been made[15]. The mean
free path λ and the length of cathode dark space Z_c
are assumed to be 0.5 cm and 10 cm, respectively,
under the typical condition. When applied voltage
V_c is 3 kV and gas pressure P is 10^{-2} torr, the
ions lose almost 90 % of the energy dissipated in
the glow discharge by transferring this energy to
neutral atoms in traversing the cathode dark space.
The energetic neutral atoms (less than 135 eV) are
in the majority and ions with moderately high energy
(about 300 eV) are in the minority for film formation,
even in the method where the substrate is set in
the plasma region and the applied voltage is 3 kV.

In contrast, ion beam deposition is the genetic
term for the film deposition in a high vacuum region ($< 10^{-4}$ torr).

The ion beam deposition methods including their
modified type are classified as follows: i) hybrid
methods involving vacuum evaporation and ions or
an ion beam and ii) the ion beam deposition using
an ion source as the main constituent. The method
i) will be classified into A) the method of simultaneous vacuum evaporation and ion implantation
or ion irradiation and B) the partially ionized

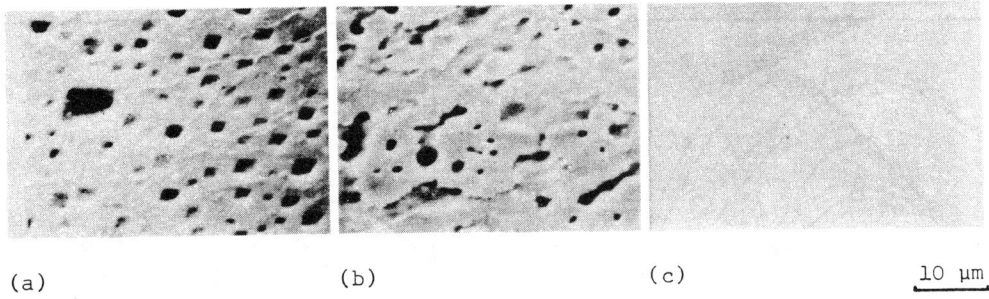

(a)　　　　　　　　(b)　　　　　　　　(c)　　　　　10 μm

Fig.5　Scanning electron micrographs of Al-Si interface. (a) ICB deposition, annealed at 450°C,
(b) vacuum deposition annealed at 450°C and
(c) ICB deposition annealed at 200 °C.

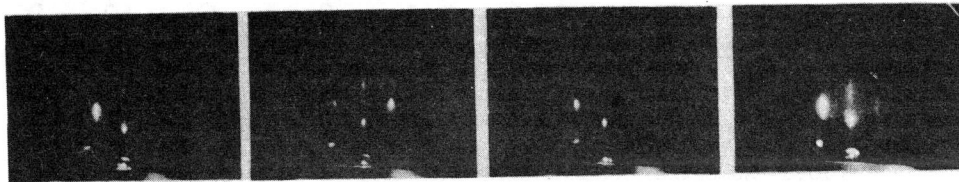

Va = 0.2 kV　　　Va = 2.0 kV　　　Va = 4.0 kV　　　Va = 6.0 kV

UNDER U.H.V. (5 X 10^{-9} torr) CONDITION

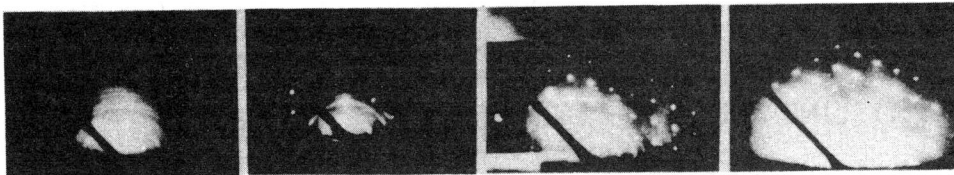

Va = 0 kV　　　Va = 4.0 kV　　　Va = 6.0 kV　　　Va = 8.0 kV

UNDER H.V. (1 X 10^{-6} torr) CONDITION

Fig.6　Electron diffraction patterns of epitaxial
silicon films by ICB at different acceleration
voltages.

vapour deposition method. In the former, an auxiliary ion source for producing metal ions or reactive gas ions is combined with a conventional evaporation method, and in the latter the source material which is partially ionized is used for film formation together with the neutral state source materials. Some of the results according to this classification are listed in previous report[1].

An ionized cluster beam deposition is a kind of ion beam deposition but is different in that, in this deposition, ionized and neutral clusters instead of atomic state are used. This will be discussed in next section.

4. Ionized cluster beam deposition
4.1. Ionized cluster beam

The ionized cluster beam (ICB) technique was originally developed by T.Takagi et al.[16] In this technique, ions of macroaggregate atoms (cluster) are utilized instead of ions in atomic state. The clusters are formed by super-saturation caused by adiabatic expansion through the nozzle without the mixture of any inert gases[17]. A cluster contains 500 - 2000 atoms loosely coupled together. The clusters are ionized by an electron bombardment. Then the cluster ions are accelerated toward the substrate by a negative high potential.

One of the most important properties of the cluster ion is elimination of problems caused by the space charge effects due to the extremely small charge-to-mass ratio. In addition, the cluster beam is well oriented at high intensity. These features make the cluster beam able to transport large mass at a very low energy. In the ICB technique, not only the advantages in such macroscopic properties but also the microscopic properties are utilized. A microscopic structure of the cluster formed by this method were studied by an in-situ electron beam diffraction analysis[18] (in case of Sb), and was found that the cluster was amorphous. The atomic distances in the cluster are 2 - 8% longer at different sites of atoms compared with the bulk crystal. This characteristic will offer unique film formation kinetics such as enhanced adatom migration.

4.2. Nucleation and growth characteristics

The energy of the ionized clusters can be controlled by applying an acceleration voltage. In the cluster ion, energy of each constituent atom is given by dividing energy of a cluster by the number of constituent atoms in the cluster. Assuming that the cluster ion is accelerated by the potential of 1 kV, the energy of constituent atom is 1 eV for a cluster of 1000 atoms. Figure 3 shows the dependence of density of nuclei as a function of deposition time at different acceleration voltages (a) and the ionization currents (b). Lead clusters were deposited onto a copper-grid covered with carbon films keeping the deposition rate constant by quartz thickness monitor. The density of nuclei is increased monotonically at first by increasing the acceleration voltage up to about 1 kV and then saturated. In addition, it was observed in another experiment[19] that the adatoms can migrate a considerable distance on the substrate surface by imported momentum until they are trapped by nuclei. Therefore we can expect that in the initial stage of film formation the kinetic energy of ionized clusters affects both the adatom migration and the density of nuclei.

Increase in the ionization ratio of the clusters causes little increase in the density of nuclei, but adatom migration could be enhanced as was shown in previous section. Other experiments of reactive ionized cluster beam deposition showed that oxide, nitride or hydrogenated films could

be formed under 10^{-5} - 10^{-4} torr of reactive gases [20]. Therefore effect of ionic charge transported by beam to the substrate surface is considered to enhance adatom migration and to generate chemically active state of reactive species.

4.3 Film growth characteristics

In conventional vacuum deposition and electroplating, films tend to grow with preferred orientation (fiber texture) on an amorphous substrate. For example, vacuum deposited films of F.C.C. metals on amorphous substrate exhibit a (111) orientation and compounds with the wurtzite structure show pronounced (0001) orientation. When cluster ions are used, the degree of preferred orientation can be easily controlled by changing the content of electric charge and the kinetic energy of the clusters.

To elucidate the relation between the acceleration voltage and crystalline structure, several examples will be shown here. Lead was deposited onto an amorphous substrate at different acceleration voltage. The result is shown in Fig. 4. The acceleration voltage is increased from 1 - 6 kV, then the intensity of (111) diffraction increases at first and reaches the maximum value. In contrast, the (200) intensity decreases with increasing acceleration voltage. Similar results were obtained in case of Ag and Au deposition[17].

Al films were also formed on single crystal Si substrates[21]. In this experiment, effective cleaning process of a native oxide on Si single crystal could be observed. After annealing of the films deposited by ICB and vacuum deposition at 450 °C for 30 min. in vacuum, Al films were chemically etched off and the Al-Si interface was observed by SEM. Figure 5 shows the clear contrast of the SEM images of the sample surfaces. In the sample deposited by the ICB, square etch pits arranged regularly are seen whereas in the sample deposited by vacuum deposition no regularity is observed. The fact indicates that in ICB deposition, substrate cleaning due to the bombardment by ions causes a uniform interface without any oxide or impurity islands. For a practical application to semiconductor metallization, contact properties and resistivity become important. In ICB deposited films, sufficiently low contact resistance was obtained after annealing at 200 °C whereas in vacuum deposited films annealing temperature up to 450 °C was necessary. As shown in Fig. 5 interface between the substrate and Al was smooth when 200 °C annealing was performed.

In the silicon epitaxial deposition, silicon film could be formed at a reasonably high deposition rate with ICB at substrate temperature below 800 °C in a chamber which was evacuated with oil diffusion pump to a base pressure of 10^{-7} - 10^{-6} torr[22]. In this deposition, special cleaning process was not used before the deposition except chemical cleaning. Diffraction patterns by 100 kV electron beam were used to evaluate the crystalline structure. Silicon epitaxial deposition was also performed in ultra high vacuum chamber by ICB. The atomically clean and well-ordered silicon surface was obtained by direct current heating of the substrate at 1250 °C at a background pressure of 5 X 10^{-10} torr and subsequently cooling down it to 500 °C. In-situ observation of diffraction pattern was made during the deposition by medium electron energy at 5 kV. Figure 6 shows the comparison of the samples formed by different background and substrate cleaning conditions. In the deposition on substrate surface covered by native oxide, polycrystalline pattern could be obseved up to 6 kV acceleration voltage. It seems to be necessary to apply high voltage to accomplish sufficient surface cleaning before epitaxial deposition. On the other hand, in the deposition on atomically clean surface in the UHV, single crystalline pattern is observed at 200 V acceleration. By increasing the acceleration voltage, improvement of the crystalline quality could be observed. This might be due to the enhanced migration effect.

These experiments show that the important fundamental film formation processes such as sputter cleaning, nuclei formation, adatom migration, coalescence and over-growth are effectively controlled by the intensity of the kinetic energy and content of cluster ion.

A large number of elemental and compound material films deposited by the ICB process have been demonstrated. Examples are summarized in recent papers [17, 20].

5. Conclusions

To study the features of ion plating and ion beam deposition, effects of kinetic energy and ionic charge in the early stages of the film formation and their relationship to subsequent film characteristics were discussed. Specific differences between ion plating and ion beam deposition are mainly due to the difference in the relative fraction of ions, energetic neutrals and thermalized gases. The range of the kinetic energy used to control physical and crystallographic characteristics of the film was investigated using experimental results. It was found that the energy of ions effective for film formation typically ranges from a few eV to several hundred eV. The effects of ionic charge were also studied and shown to influence adatom migration, coalescence and chemical activity of adatoms.

Fundamental properties of films deposited by ICB were also discussed. ICB deposition technique realizes high intensity low energy ion beam and the enhancement of adatom migration by ICB is especially effective in improving the film quality.

REFERENCES

1. T.Takagi: Thin Solid Films, 1982,92, to be published.
2. J.E. Greene and S.A. Barnett: J.Vac. Sci. Technol., 1982, July/Aug. issue to be published.
3. G.Carter and D.G. Armour: Thin Solid Films,1981, 80, 13-29.
4. N.Matsunami: Proc. 4th Symp. on Ion Sources and Ion-Assisted Technol, Tokyo,1981, Institute of Electrical Engineers of Japan, Tokyo, 1981,315-318.
5. G.H. Kinchin and R.S.Pease: Rep. Progr. Phys., 1955,18,1-51.
6. K.L.Chopra: J. Appl. Phys.,1966,37,2249-2254.
7. Y.Namba and T.Mori:J.Vac. Sci. Technol.,1976,13, 693-697.
8. Y.Namba and T.Mori: Thin Solid Films,1976,39, 119-123.
9. T.Takagi, I.Yamada and K.Matsubara: Thin Solid Films, 1979,58,9-19.
10. T.Takagi, K.Matsubara and H.Takaoka: J. Appl. Phys.,1980.51,5419-5424.
11. T.Spalvins: J. Vac. Sci. Technol.,1980,17,315-321.
12. Y.Murayama: J.Vac. Sci. Technol., 1975,12,876-883.
13. M.Tatthews and D.G. Teer: Thin Solid Films, 1981, 80,41-48.
14. R.Adach and K.Takeshita: J.Vac. Sci. Technol., 1982, 20, 98-99.
15. D.G.Teer: Proc. Ion Plating & Allied Techniques, IPAT'77, Edinburgh, 1977. CEP Consultants, Edinburgh, 1977,13-31.
16. T.Takagi, I.Yamada, M.Kunori and S.Kobiyama: Proc. 2nd Int. Conf. on Ion Sources, Vienna, 1972, Osterreichishe Studien-Gesellshaft für Atomenergie, Vienna, 1972, 790-795.
17. I.Yamada and T.Takagi: Thin Solid Films, 1981, 80, 105-115.
18. I.Yamada, G.D. Stein, H.Usui and T.Takagi: Proc. 6th Symp. on Ion Sources and Ion-assisted Technol. Tokyo, 1982, Institute of Electrical Engineers of Japan, Tokyo, 1982, 47-52.
19. T.Takagi, I.Yamada and A.Sasaki: Thin Sold Films, 1976, 39,207-217.
20. T.Takagi, I.Yamada and H.Takaoka: Surface Science, 1981, 106,544-550.
21. H.Inokawa, K.Fukushima, I.Yamada and T.Takagi: Proc. 6th Symp. on Ion Sources and Ion-Assisted Technol., Tokyo, 1982, Institute of Electrical Engineers of Japan, Tokyo, 1982,355-358.
22. I.Yamada, F.W. Saris, T.Takagi, K.Matsubara, H. Takaoka and S.Ishiyama, Japan. J. Appl. Phys., 1980,19,181-184.

H ISHIMARU

Chromium-nitride coating on aluminum alloy vacuum flange by ion plating

SYNOPSIS

Chromium-nitride coating by an ion-plating method
was carried out on the surface of an aluminum
alloy vacuum flange. The thickness of the CrN
layer was about 3 µm. The surface hardness of the
CrN was very high and its micro-Vickers (100 g)
was about 1800. The CrN treatment on the aluminum
alloy flange gave nearly perfect protection against
sticking between the knife edge of the flange and
the gasket, and against surface scratching. This
flange system is compatible with the ordinary
stainless steel Conflat flange in ultrahigh vacuum.

THE AUTHOR

is in the National Laboratory for High Energy
Physics, Ibaraki-ken, Japan.

INTRODUCTION

Aluminum and aluminum alloys are preferred materials
for the vacuum chambers of electron storage accelerators because of their good thermal conductivity,
extremely low outgassing rate, low residual radio-
activity, and the fact that they are completely
non-magnetic. The vacuum chambers currently used
are of aluminum with conventional stainless steel
Conflat flanges and bellows in PETRA,[1] PEP,[2] and
KEK-PF.[3] An aluminum--stainless steel transition
has been used in the construction of these machines.
This Al--SS transition has low reliability, high
cost, low thermal conductivity, complicated
structure, high residual radioactivies and high
outgassing rate for mild bakeout. The bakeout
temperature of an aluminum alloy system is limited
to approximately 150°C. If ordinary stainless steel
components are mixed in an otherwise aluminum alloy
system effective bakeout for the stainless steel
is impossible because of the limitation due to the
aluminum. Then the ultimate pressure is limited by
outgassing of the stainless steel components.
Ordinary stainless steel and the stainless steel--
aluminum transition system have been eliminated as
far as possible from the TRISTAN design.[4]

ALUMINUM ALLOY CONFLAT FLANGE

A flange system made entirely of aluminum alloy
was developed.[5--7] The basic feature of the system
is the use of 2219-T87 aluminum alloy, which has
the same mechanical properties as stainless steel,
for the flange material and an aluminum alloy
gasket or Helicoflex. Alloy 2219-T87 has the highest
strength at elevated temperature of all aluminum
alloys as well as superior weldability and stress
corrosion cracking resistivity. Therefore, it can
be considered to be the best all-aluminum alloy
for use as a flange material for ultrahigh vacuum
equipment that requires bakeout. The aluminum
alloy flange resembles the traditional Conflat
types (Fig. 1). The knife edge part is given a
super-finished surface[8] processed by a flat diamond
tool. Anodized aluminum alloy bolts (2024-T4), non-
anodized nuts (6063-T6) and a hard anodized washer
(2017-T4) are used to tighten the flange.

CrN COATING

The special aluminum alloy (2219-T87) is suitable
for using an aluminum gasket and a Helicoflex O-
ring but the aluminum surface has poor resistance
to abrasion. The sealing performance depends
largely on the delicate condition of the sealing
surface of the flange. Sticking between the knife
edge of the flange and the gasket is observed

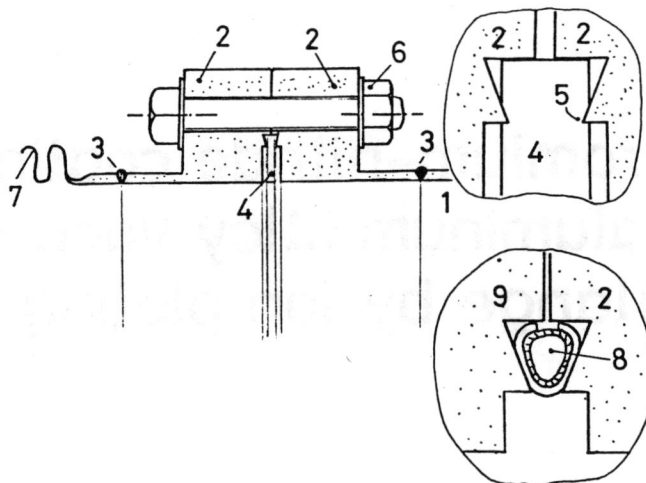

Fig. 1 Aluminum alloy Conflat type flange/gasket/bolt, nut and washer combination
  1 aluminum alloy pipe (6063-T6)
  2 aluminum alloy flange (2219-T87)
  3 AC-TIG welding
  4 aluminum gasket (3004-H34)
  5 CrN coating
  6 anodized aluminum alloy bolt (2024-T4), non-anodized nut (6063-T6),
     and hard anodized washer (2017-T4)
  7 aluminum alloy corrugated bellows (3004)
  8 Helicoflex O-ring
  9 stainless steel Conflat flange

during the bakeout. The CrN treatment on the flange, which gave nearly perfect protection against sticking and surface scratching, was carried out by the ion-plating method.[9]

TESTING CrN COATING

Separation between the CrN film and the 1 mm thick aluminum alloy specimen was not observed even after bending at right angles. No penetration of the red color check liquid into the CrN film was observed. No microcracks were detected on the CrN film. The surface hardness of the CrN was very high and its micro-Vickers value (100 g) was about 1800 with a thickness of more than 3 μm. For reliable sealing, the necessary superfinished surface can be achieved by using a flat diamond tool. The bonding between the CrN film and both flanges was favorable. The necessary lathe work on the CrN coated aluminum flange was performed before welding. Separation between CrN film and aluminum alloy flange could not be observed after welding.[10] No change was noticed on the CrN surface after cycling of the $150°C$, 24 hours bakeout procedure with vacuum. When the CrN coated aluminum alloy flange was bombarded by electron beam[11] (60 MeV, 40 μA and 5 mmφ), the temperature increased to $200°C$ during 20 min without cooling but no change was noticed on the CrN surface. No change of color was observed.

ELECTRON DIFFRACTION

The electron diffraction pattern of the CrN film[12] revealed that this film was the crystal of the CrN. A small quantity of CrN_2 was also observed.

CONCLUSION

As described above, a complete surface treatment for its aluminum alloy vacuum flange has been developed for the TRISTAN electron-positron collider.

ACKNOWLEDGMENTS

The author wishes to thank Tigold Corporation[13] for their work on the CrN coating of the aluminum alloy flange. Acknowledgment is due to Prof. I. Yoshizawa, Ibaraki University, for taking a photograph of the CrN electron diffraction pattern.

REFERENCES

1. H. Hartwig and J. Kouptsidis: Proc. 7th Inter. Vac. Congress, Vienna, 1977.
2. J.R. Rees: IEEE Trans. NS-24, 1977.
3. M. Kobayashi et al.: Nuclear Instrum & Method, 117 (1980).
4. H. Ishimaru, G. Horikoshi and Y. Kimura: IEEE Trans. NS-28, No. 3, 1981, 3320.
5. H. Ishimaru: Journal Vac. Sci. & Technol., 15, 1978.
6. H. Ishimaru, G. Horikoshi and K. Minoda: IEEE Trans. NS-26, No. 3, 1979.
7. H. Ishimaru and S. Shibata: 3rd Symp. Acc. Sci. Technol., Aug. 1980, Osaka Univ.
8. I. Sakai, H. Ishimaru and G. Horikoshi: Vacuum, Vol. 32, No. 1, 1982, 33.
9. S. Komiya, N. Umezu and C. Hayashi: 6th International Vacuum Metallurgy Conf. on Special Meeting and Metallurgical Coatings, 1979, San Diego, U.S.A.
10. K. Tsuchiya et al.: IHI Engineering, Rev. Vol. 14, No. 3, 1981, 21.
11. H. Ishimaru and O. Konno: Research Report of Lab. of Nuclear Sci., Tohoku Univ., 1981, 113. (in Japanese)
12. H. Ishimaru: Workshop on First Wall Coating, IPPJ-551, Mar. 1982, 219.
13. Tigold Corporation, 516 Yokota, Sanbu-machi, Chiba, 289-12, Japan.

H BARTON

Ivadizing: ion vapour deposition of aluminium

SYNOPSIS

The worldwide call for a reduction in the use of cadmium on environmental grounds and the problems associated with cadmium on high strength steels, aluminium alloys and titanium were major reasons for extensive investigations into the alternative methods of corrosion protection.

Ivadizing has shown itself in service to be an acceptable alternative to cadmium and because of its performance advantages can be used in a wide range of applications.

THE AUTHOR

is Vacuum Plant Sales Manager at General Engineering Radcliffe 1979 Ltd.

Introduction

For many years in the aerospace industry cadmium electro-plating coatings have been the most common method of corrosion protecting of steel components.

However, as the use of high strength aluminium and steel alloys became more widespread problems arose with the use of cadmium.

On high strength steels electroplated cadmium often caused hydrogen embrittlement, necessitating an additional de-embrittlement treatment. On titanium cadmium is a cause of solid metal embrittlement. When cadmium plated fasteners are installed in high strength alloys exfoliation corrosion occurs.

In the early 1960's the McDonnell Douglas Corporation began to look for a viable alternative to cadmium whilst at the same time they investigated vacuum deposition of cadmium as a solution to hydrogen embrittlement, having considerable success.

The use of aluminium in a vacuum environment was investigated and deposition by the ion vapour method was found to give the best results with regard to adhesion, uniformity of coating and corrosion resistance.

After many years of development, production machines are now being manufactured and marketed under the registered name of IVADIZER.

The Process

Ivadizing (Ion Vapour Deposition of Aluminium) is a vacuum process similar to the more common vacuum metallising process. However, with Ivadizing the workpiece is given a high negative potential with respect to the evaporation source.

The equipment consists of:

a. A vacuum chamber.

b. A vacuum diffusion pumping system.

c. An aluminium evaporation source.

d. A high voltage power unit.

e. A gas control system.

(See Schematic Figure 1)

The treatment cycle consists of:

1. Evacuation of vessel.

2. Backfill with argon to control pressure.

3. Apply high voltage and glow discharge clean.

4. Coat.

5. Backfill to atmospheric pressure and remove parts.

Advantages of Ivadizing

Ivadizing has several advantages over other protective coatings used on steel and aluminium alloy parts.

These include:

1. Usable temperature up to 496°C compared to 232°C for cadmium.

2. IVD can replace diffused nickel/cadmium giving better corrosion protection at all strength levels. Diffused Ni-Cad is limited to steels having strength levels below 200,000 lb/sq.inch. due to hydrogen embrittlement.

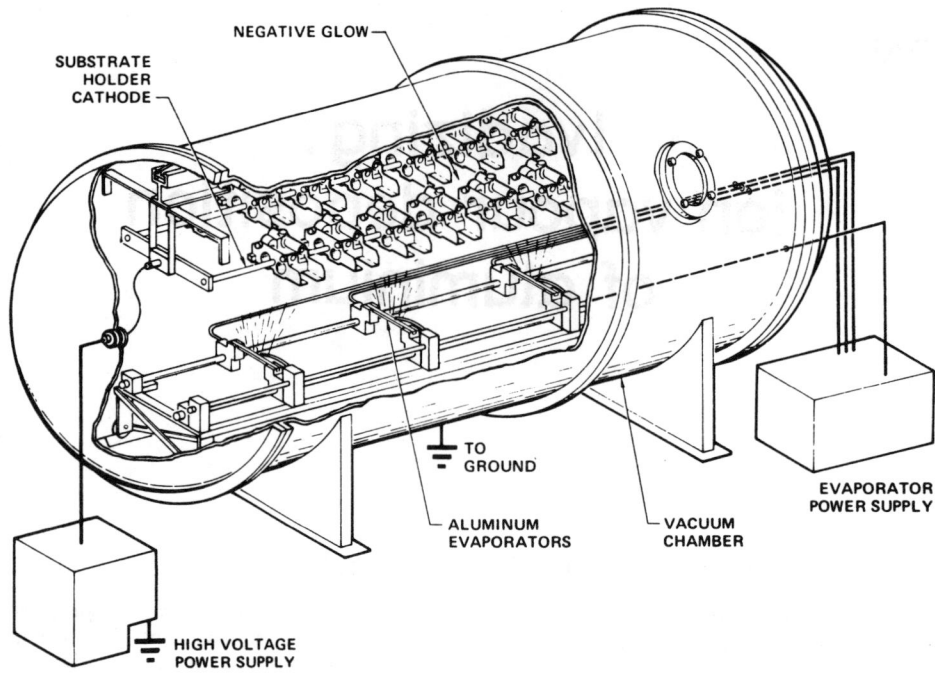

FIGURE 1
SCHEMATIC OF AN ION VAPOR DEPOSITION SYSTEM

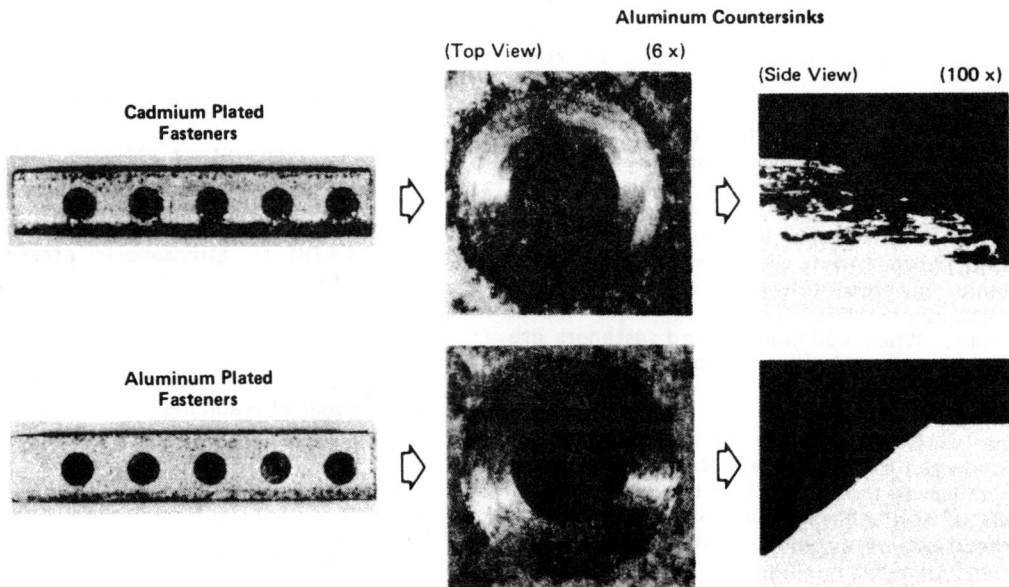

FIGURE 2
IVD ALUMINUM AND CADMIUM COATED STEEL FASTENERS
INSTALLED IN 7075-T6 ALUMINUM ALLOY AND EXPOSED
TO 168 HOURS OF SO_2 - SALT SPRAY

3. IVD does not cause solid metal embrittlement of titanium. Cadmium plating is prohibited.

4. IVD can be used in contact with fuel. Cadmium is prohibited.

5. IVD provides galvanic protection to aluminium alloys and does not cause fatigue reduction. Anodised coatings provide only barrier coating protection and cause fatigue reduction.

6. IVD can be applied thinner than alclad coating on aluminium alloys resulting in weight savings and is not limited to rolled forms.

7. IVD does not cause any ecological problems.

Coating Performance

Three classes and two types of coating are used.

The classes relate to the thickness of coating whilst the type are I, as coated and II chromated. Chromating adds to the corrosion resistance and provides a suitable surface for painting.

U.S. Military specification MIL-C-83488B calls for corrosion resistance as Table 1 when subjected to a salt spray test to ASTM method B-117.

Class	Thickness	Test Period	
		Type I	Type II
1	25 micron (min)	504	672
2	12.5 micron (min)	336	504
3	7.5 micron (min)	168	336

Class 1 coatings are used in severe corrosion environments.

Class 2 coatings are used in less severe interior corrosion environments.

Class 3 coatings are used when fine tolerances are necessary on such parts as fine threaded parts.

Coating uniformities and adhesion properties are comparable to electroplating. Figure 4 shows and example of coating uniformity on a fastener.

Numerous corrosion tests have been performed by various concerns. Results have been wide ranging but using cadmium as comparison various conclusions can be used.

When cadmium electroplated and IVD aluminium steel parts are subjected to a 5% salt spray cadmium generally gives better protection, however, should a scratch be made through the coatings cadmium tends to sacrifice itself quicker and allows red dust to appear before IVD aluminium.

Cadmium and IVD aluminium coated steel fasteners installed in aluminium alloys have been exposed to both 5% neutral salt fog and also SO_2 - salt spray environments. After 2500 hours exposure to the neutral salt fog environment the IVD coated fastener heads were more corroded than the cadmium plated fasteners. However, the countersinks where the IVD coated fasteners were installed were not nearly as corroded as those countersinks in which the cadmium fasteners were installed.

After 168 hours in the SO_2 — salt fog environment both the heads of the cadmium plated fasteners and the countersinks into which they were installed were much more severely corroded than the IVD aluminium ones (Figure 2).

IVD coated titanium fasteners and bare titanium fasteners installed with wet epoxy primer have been installed in an aluminium structure and exposed to SO_2 - salt spray. After 24 days examination showed blistering around the peripheries of the fasteners installed with wet primer which was more severe around the IVD aluminium coated fasteners. Examination of the countersinks showed more corrosion where the cadmium plated fasteners were installed. (See Figure 3).

Production Status

As early as 1974 production machines have been manufactured and IVD aluminium coated parts have been used in service.

Examples of the use of IVD (Figure 7) are to be found on the F-15 Eagle, the F-18 Hornet, AV-8B Advanced Harrier, DC-9, DC-10, F-4 retrofit and harpoon missile. On future McDonnell Douglas designs IVD will be used extensively.

Over thirty production units have been produced to date and supplied to both the aerospace industry and sub-contract platers. The majority have been in the U.S.A. but more recently units have been sold in Japan, France and U.K.

Three basic types of machine have been developed (See Figure 2) which cater for most components shapes.

1. The Rack Coater (Figure 5) in which large components are hung from a rack and are static whilst the evaporation source translates.

2. The Carousel Coater in which components are held on two planetary fixtures.

3. The Barrel Coater (Figure 6) for fasteners and small components.

   In this machine two mesh barrels rotating over the evaporative source form the support jig. Inlet and outlet hoppers are provided to provide a semi-continuous operation.

In late 1980 a license was granted to General Engineering Radcliffe 1979 Limited to manufacture and market the Ivadizer range of equipment in Europe and Scandinavia.

General Engineering has for many years been manufacturing a wide range of vacuum plant as well as its own range of vacuum pumps. The Ivadizing machines dovetail into the General Engineering range very well since:

1. Ivadizing is a vacuum process requiring a vessel and vacuum pumping system.

2. Ivadizing utilises a wire fed resistance heated boat type of evaporation source which is similar in many ways to the system used on the General Engineering Roll Coater Metallisers.

IVD Aluminum Coated
Fastener

Bare Titanium Fastener
Wet Installed

GP77-0982-14

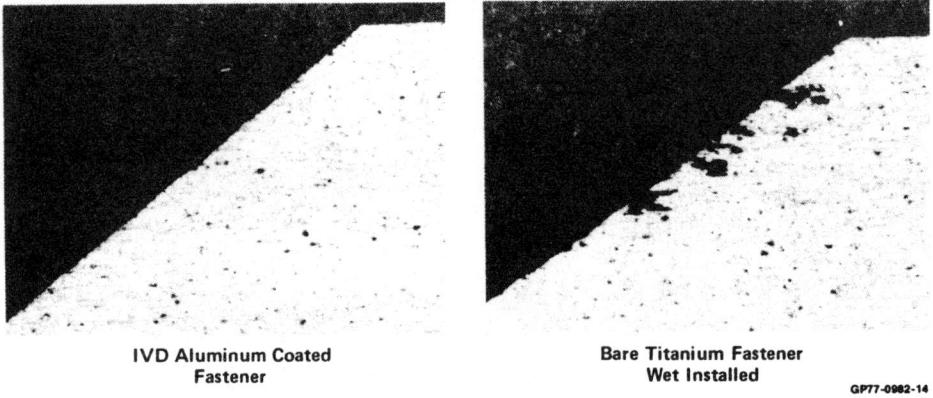

**FIGURE 3
CORROSION TESTS OF IVD COATED AND WET INSTALLED TITANIUM FASTENERS
IN 7075-T6 ALUMINUM ALLOY COUNTERSINKS**

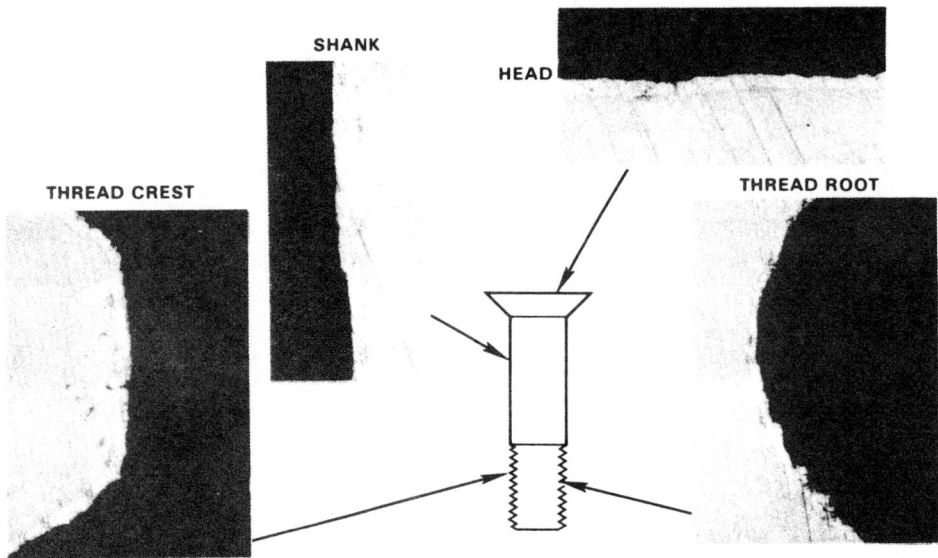

SHANK

HEAD

THREAD CREST

THREAD ROOT

**FIGURE 4
IVD ALUMINUM THICKNESS DISTRIBUTION ON A FASTENER**

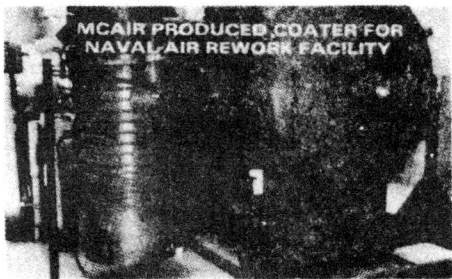

Naval Air Rework Coater

Barrel Coater

McDonnell Production Coater

Air Force Coater

GP77-0982-15

**FIGURE 5
PRODUCTION COATERS**

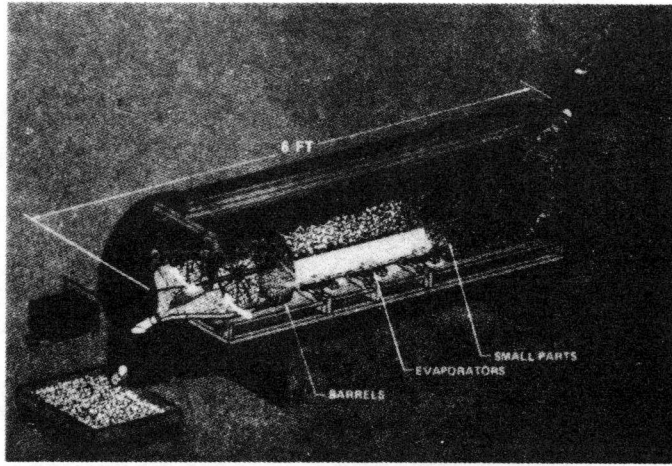

FIGURE 6
SCHEMATIC OF BARREL COATER

Landing Gear

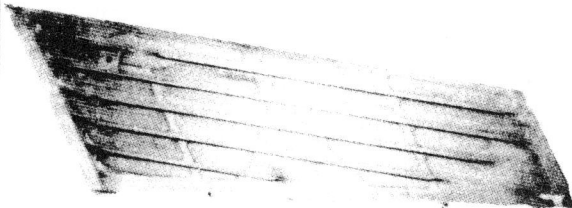

Wing Skin

Engine Mount

Bellcrank

Stator Vane Assemblies

FIGURE 7
IVD ALUMINUM COATED AIRCRAFT AND ENGINE PARTS

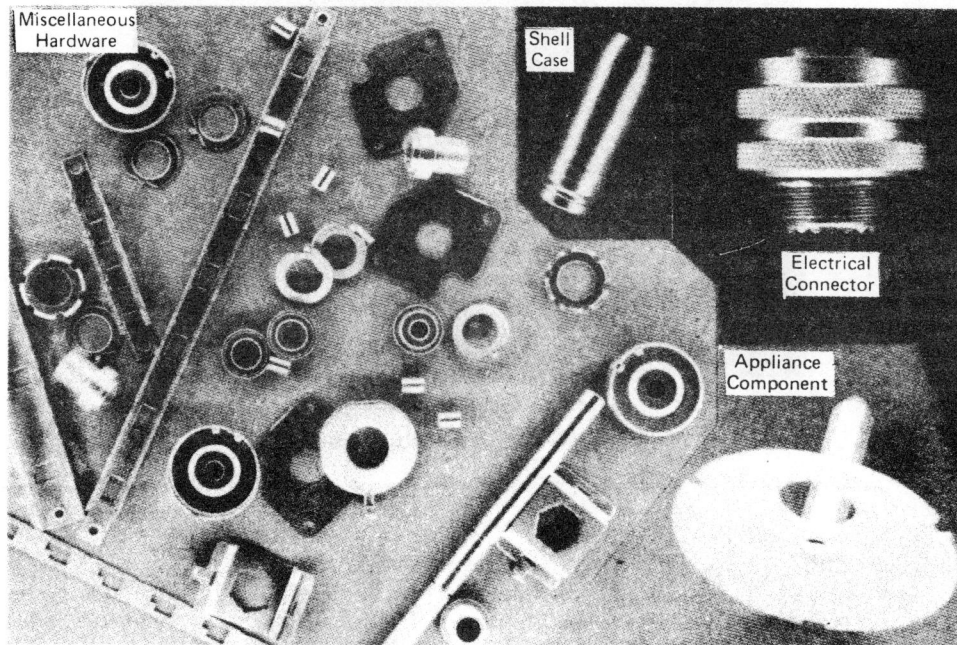

FIGURE 8
IVD ALUMINUM COATED PARTS

3. The high voltage glow discharge and argon gas control system used to produce the plasma cleaning operation is similar to that used on the General Engineering Sputtering Machines.

Much interest has been shown in Ivadizing particularly from the aerospace, nuclear, ordnance and automobile industries. This interest is aided by the worldwide wish to reduce the use of cadmium.

The U.S. Department of Defence now recognise IVD aluminium as an alternative to cadmium.

Airbus Industries are now using IVD aluminium on titanium fasteners.

Economics of the System

With regard to operating costs the capital cost of the plant is the major factor followed by labour costs and then materials and utilities.

When comparing costs against electroplated one must take into account secondary costs such as effluent treatment, post heat treatments like de-embrittlement and also the fact that Ivadizing can be used to replace a number of processes e.g. NiCd, Cd, anodising and Cermetal.

Approximate capital costs of Ivadizers are:

a.	4' x 6' Carousel Coater	£105,000
b.	4' x 6' Barrel Coater	£120,000
c.	4' x 6' Rack Coater	£140,000
d.	6' x 10' Rack Coater	£200,000

Labour requirement is one man or less per machine.

Cycle times vary between 20 minutes and $1\frac{1}{2}$ hours dependent on loads and materials to be coated. The plating capacity of a 4 foot diameter by 6 feet long barrel coater is approximately 120 pounds per hour whilst the capacity of the rack coaters is dependent on the size and shape of the components.

The costs of materials and utilities are approximately £3.00 per hour for the barrel and small rack coaters and £6.00 per hour for the larger rack coater.

Conclusions

Ion vapour deposited aluminium coating has now shown through a number of years of service to be a high performance protective coating and an acceptable alternative to cadmium plating. The coating has been shown to have excellent properties of corrosion protection, uniformity of coating and adhesion. It overcomes the problems of hydrogen embrittlement of high strength steel alloys and solid metal embrittlement of titanium alloy without contributing to the pollution of the environment.

3.6

A MATTHEWS

Avoidance of excessive substrate temperatures during plasma assisted processes

SYNOPSIS

A thermodynamic analysis is outlined which permits the prediction of specimen temperatures during plasma assisted processes. This analysis can be used to ensure that optimum temperatures are achieved and substrate deterioration is avoided. The paper also gives guidance on the effects of specimen size and shape variations on the uniformity of bombardment. The utilisation of discharge support techniques, which allow the bias voltage, current and pressure to be controlled independently and also enable the nature of the discharge to be changed, is seen as an important development in process control.

THE AUTHOR

Dr Matthews, formerly at the University of Salford, is now with the Department of Engineering Design and Manufacture at the University of Hull.

INTRODUCTION

Plasma assisted processes lead to substrate heating through ion and neutral bombardment. In most cases this heating is advantageous, improving coating adhesion and structure properties in ion plating[1,2] and aiding diffusion in plasma assisted surface treatments. However, it is important, when using these processes, to ensure that the level of substrate heating produced is optimal; most importantly heating should not be deleterious to the bulk substrate properties, as occurs when the tempering temperature of hardened steel substrates is exceeded.

Continuous monitoring of the sample temperature during bombardment, by methods such as optical pyrometry or thermocouple techniques, presents severe difficulties. Readings by the former method are influenced by the discharge glow, and the latter technique requires that a suitable means be found of attaching the thermocouple wires to the sample. Also if the wires are not shielded they may be eroded by bombardment, or the reading may be influenced by transient arcing. In any event such monitoring techniques may only detect a process error after the damage has occurred. It is far more desirable to have some means of predicting specimen temperatures in advance so that the process can be pre-planned and controlled much more effectively.

There follows a thermodynamic analysis which can be used in such a way to predict temperatures achieved by a sample which is subject to ion and neutral bombardment as the cathode of a glow discharge. Subsequently the value of this analysis in some typical practical applications will be discussed.

THE THERMODYNAMIC ANALYSIS

In any low pressure plasma assisted process the discharge power input to the sample is dissipated principally by thermal radiation to the chamber walls and by heat conduction through the cathode. Conduction and convection effects through the partial vacuum will be negligible. We may designate the major heat flows as shown in Fig 1.

Q_E is the discharge power input to the sample, carried by ions and accelerated neutrals.

Q_{R13} is the thermal radiation transferred from the specimen to its surroundings.

Q_C is the heat conducted from the specimen to the cooled cathode.

Q_{R21} designates heat radiation transferred to the sample from a vapour source, when this is present.

During evaporation there will also be a heat input through condensation, as well as thermal radiation from the vapour source. In many applications the former will be negligible compared to the latter. In the case of titanium deposited at $1\mu m\ min^{-1}$ for example the heat of condensation amounts to about 3×10^{-3} W cm^{-2}. Discharge power inputs typically exceed 1 W cm^{-2}. The following balance will in general therefore apply when steady state conditions are reached:

$$Q_{R21} + Q_E = Q_{R13} + Q_C \qquad (1)$$

In non-evaporative plasma assisted processes the Q_{R21} term will not be present.

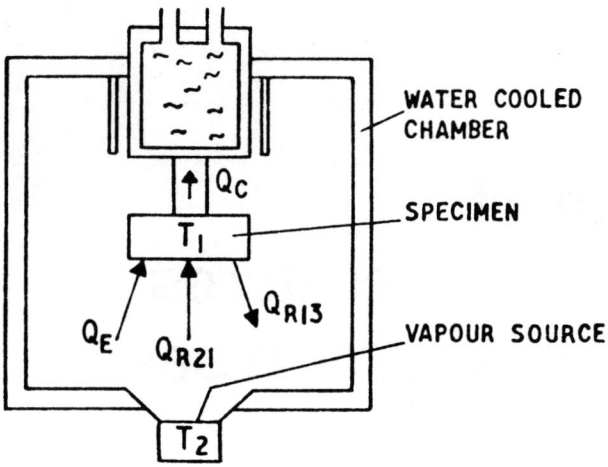

1. Simplified heat transfer diagram.

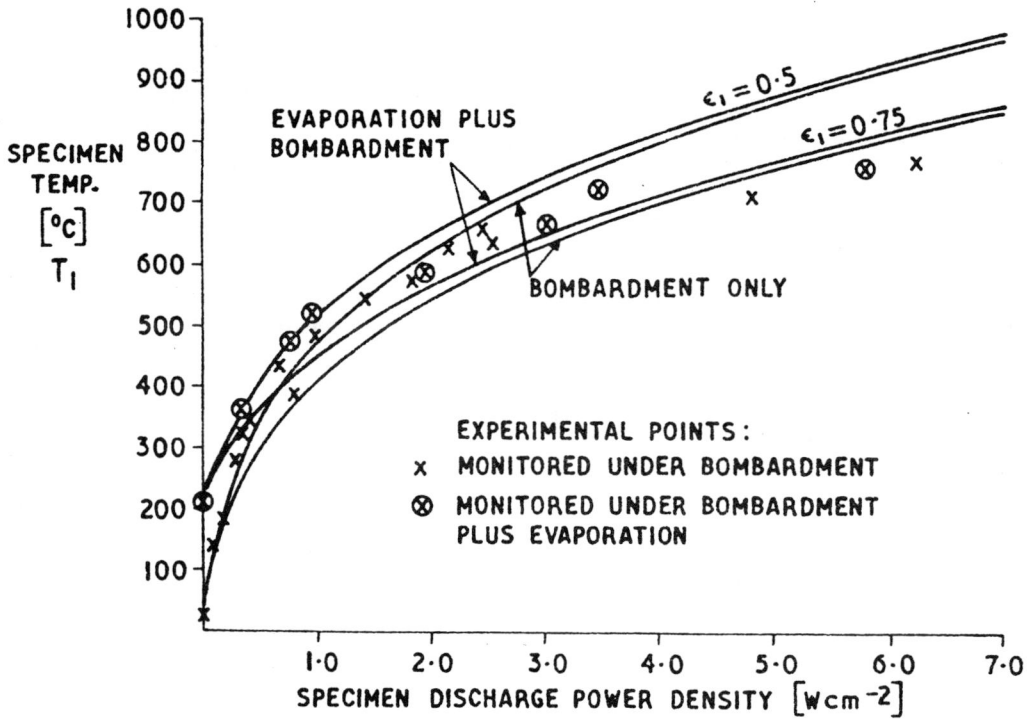

2. Temperature prediction curves produced using
    the thermodynamic analysis.

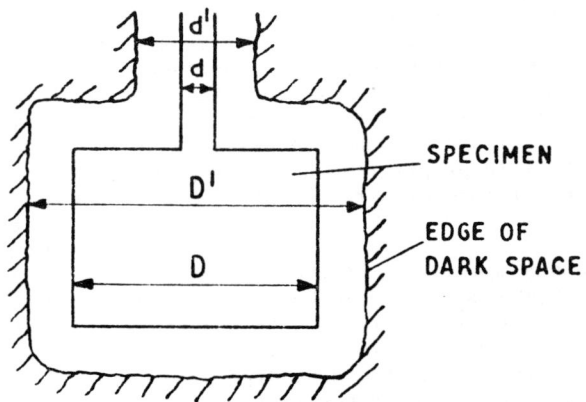

3. Figure to illustrate the effect of the glow on
    uniformity of bombardment.

In an earlier paper[3] the author outlined how thermal radiation theory may be applied to assess Q_{R21} and Q_{R13}. By using simplifying assumptions it was shown that

$$Q_{R21} = A_1 \varepsilon_1 \varepsilon_2 \sigma (T_2^4 - T_1^4) \ F_{\overline{1-2}}$$

and $$Q_{R13} = A_s \ F_{\overline{1-3}} \ \sigma \ \varepsilon_1 \ (T_1^4 - T_3^4)$$

A_1 is the specimen area viewed by the vapour source.
A_s is the total specimen area.
ε_1 and ε_2 are the emissivities of the specimen and the vapour source, respectively.
σ is the Stefan-Boltzmann constant.
T_1 is the specimen temperature, T_2 the surface temperature of the vapour source and T_3 is the temperature of the chamber walls.
$F_{\overline{1-2}}$ and $F_{\overline{1-3}}$ are heat radiation view factors from the specimen to the source and from the specimen to the chamber walls. $F_{\overline{1-3}}$ is approximately unity and $F_{\overline{1-2}}$ can be assessed by a simple geometric construction[3].

For most plasma processing applications the total exposed cathode surface area will be greater than the sample surface area, and the monitored cathode current must be proportioned so as to give the actual specimen current. This current multiplied by the specimen bias gives the maximum value for Q_E, the discharge power input to the sample. In practice the actual power reaching the sample may differ from this, for several reasons. Firstly the measured current constitutes a secondary electron current leaving the sample and an arriving ion current. It is the magnitude of the latter which we require; this will be much greater than the former. It should not of course be inferred that only ions provide the power input; neutral bombardment is the dominant source of heating[4]. Many of these neutrals may have been ions, which have undergone charge exchange collisions. Indeed, several such exchanges and impacts may occur prior to arrival at the sample. However, if we make the assumption that there is no net energy loss in these impacts then the maximum power transferred to the cathode will be given by multiplying the number of ions diffusing into the cathode fall region times the fall potential. Some energy may be lost due to energetic neutrals diffusing to the chamber walls. This proportion is unknown and it will in fact depend on the pressure and the nature of the discharge. However experiments have shown that at the same discharge power input, but across a wide range of chamber pressures, the sample achieves the same temperature; thus we will assume here that heat losses by energetic neutral diffusion to the chamber walls are negligible. Q_E is therefore taken to be the cathode bias voltage times the cathode current multiplied by a fraction to indicate the proportion going to the sample.

Q_c, the heat lost from the sample to the cooled cathode by conduction, is given by:

$$Q_c = \frac{A_H \ k \ (T_1 - T_3)}{H_H}$$

where A_H is the cross sectional area of the holder connecting the sample to the cooled cathode and H_H its length; k is the thermal conductivity of the holder.

EXPERIMENTAL DETAILS

Using the equations outlined above it is possible to predict the ultimate specimen temperature for most configurations likely to be met in plasma assisted processes. For example, if we consider a sample having a surface area of $37.5 \ cm^2$ attached to a cooled cathode by a rod 0.4 cm in diameter and 5.0 cm long, the predicted specimen temperatures at different power inputs are as shown in Fig.2. Curves are shown both for discharge heating and also for the case where this is combined with radiant heating from a vapour source, as in ion plating.

It will be noted that the temperature predicted depends on the value assumed for the specimen emissivity. As can be seen, an ε_1 of 0.5 gives the best agreement with the experimentally derived points at low power outputs, but as the power input and temperature increases a higher value of ε_1 is required. It is well known that emissivity increases with sample temperature and this must be taken into account when using this analysis technique. It should be stated though that the values of ε_1 used are somewhat higher than normally quoted at the temperatures given. The fact that the radiant heat loss from the sample is higher than anticipated is more probably due to the effective surface temperature being higher than T_1 which was measured as the bulk temperature.

It is difficult to define exactly what is meant by surface temperature of a sample which experiences ion bombardment. The arriving ions for example will have kinetic energies which may be converted to an "equivalent" temperature of several thousand degrees Centigrade. Such temperatures exist only as thermal spikes over a few atomic diameters and have no real meaning in terms of substrate deterioration or in defining a "true" surface temperature. In practice it is easier to make an assumption that the bulk and surface temperatures are the same and to modify the emissivity factors to suit this. As will be indicated later, the analysis presented here can be used to assess the effective surface temperature, though for practical purposes this is not so important as the substrate temperature.

The graph in Fig.2, whilst constructed for a particular size of sample, is in fact applicable to a wide range of sample sizes. This is because radiation is the main heat dissipation mechanism and any increase in the bombarded area of course increases the radiating area. The general applicability of the graph is confirmed by the work of Marciniak and Karpinski[5]. Their graphical results of monitored cathode temperature versus discharge power for a cathode of $1300 \ cm^2$ surface area show good agreement with Fig.2.

The curves for heating under plasma assisted deposition show several interesting features. When bombardment is accompanied by evaporation the discharge power must be reduced by an amount corresponding to Q_{R21} for the same sample temperature to be maintained. The magnitude of this heat radiation input at any temperature can thus readily be obtained from the separation of either of the two pairs of curves in Fig.2. Q_{R21} reduces as the specimen temperature increases, but its variation is small compared to that of Q_E or Q_{R13}. Whilst these two may change by an order of magnitude, Q_{R21} does not change by more than 25%, even when the specimen temperature increases by 1000°C. As an example of this effect, if we consider the evaporation of titanium at 2000°C, Q_{R21} will reduce from 10.1 W to 8.2 W for T_1 between 300°C and 1000°C. Q_{R13} would increase from 16.0 W to 487.4 W, Q_c from 2.8 W to 9.8 W, and the required Q_E to achieve these temperatures would increase from 18.8 W to 497.7 W (examples quoted for $\varepsilon_1 = 0.75$, $A_s = 37.5$ cm^2, $A_1 = 16.0$ cm^2). We can see then that the analysis can be used to determine the relative importance of the various heat sources and sinks in plasma assisted processes, and how this varies depending on the specimen temperature. The intercept with the vertical axis of the curves for simultaneous deposition gives the temperature achieved due to radiant heating from the melt only (i.e. at the zero discharge power input condition). This provides a means of assessing ε_2. The specimen temperature can be monitored readily under radiant heating only, with no discharge. ε_2 can thus be determined using equation (1) with $Q_E = 0$. In the case of titanium evaporation, for the sample size mentioned earlier, the specimen reached a temperature of 218°C. It should be noted that the temperature achieved through radiant heating will depend on the substrate geometry, as discussed in the next section.

By using "realistic" ε_1 and ε_2 values rather than those which compensate for the use of bulk rather than surface temperature, it is possible to estimate the "effective" surface temperature by using this thermodynamic analysis. The bulk temperature would be used for T_1 in the expression for Q_c, but an unknown surface temperature used in the Q_{R21} and Q_{R13} terms. By this method it was deduced that the effective surface temperature of a sample at a bulk temperature of 550°C for example exceeded 850°C. This figure must be treated with extreme caution, relying as it does on several unquantifiable variables; however, it is an interesting result, and one which may go some way to explaining the superior structural characteristics of ion plated films. From the point of view of the practical application of plasma assisted processes, however, it must be re-emphasised that it is the bulk temperature which influences substrate properties and the "effective" surface temperature remains a rather abstract concept, as currently we have no means of measuring it.

PRACTICAL CONSIDERATIONS

The author has found the preceding analysis technique to be of considerable benefit in several plasma assisted processes, particularly in his

work on the development of reactive ion plating methods for hard ceramic films, such as titanium nitride. In this process the substrate temperature can significantly affect the film properties. He therefore uses the ion bombardment or sputter cleaning stage of the ion plating process to pre-heat the sample to the desired temperature, and the above analysis is applied to predict the bombardment needed for any particular specimen geometry. The advantage of using discharge heating of the sample compared to other methods, such as direct radiant heating[6], are that the heat input is applied only where needed, i.e. to the cathode surface. Additionally there is no need to change the location of heaters for different specimens. Also this type of heating is more efficient in process time, being combined with the sputter cleaning stage. Precise details of the whole process cannot be given here. However, there follow examples of how the thermodynamic analysis would be applied to two typical substrate geometries to assess the required power inputs for pre-heat and deposition.

Consider first a rod mounted, with its axis vertical, immediately above a vapour source. This may for example represent a twist drill or similar object.

We will assume a length of 10.0 cm and a diameter of 0.5 cm. In this case the value of A_1 in the expression for Q_{R21} is 0.2 cm^2 and typical values for Q_{R21} are 0.13 W at $T_1 = 300$°C and 0.10 W at $T_1 = 1000$°C (for $\varepsilon_1 = 0.75$). These levels of power are negligible compared with typical discharge power inputs, and in this case the Q_{R21} term can be eliminated from equation (1). A further consideration with this type of sample is that clamping to ensure a sufficient conduction path to the electrode can be difficult. It may therefore be necessary to assume negligible heat conduction from the sample. Completing the analysis on this basis reveals that at a discharge power density of 1 W cm^{-2} the specimen would achieve a temperature of 500°C.

Consider next a specimen resembling a solid cylinder, held with its axis vertical. This could be a milling cutter or similar object. Assuming the diameter is 10.0cm and the length 12.0cm, in this case the power input due to radiation from the vapour source will be 49.3 W at $T_1 = 300$°C and 40.2 W at 1000°C. Though these powers are considerably higher than for the rod example, it should be noted that the cylinder has a much greater surface area than the rod and when converted to a power density over the whole surface these figures are still quite low compared to typical discharge power inputs. However, they are not sufficiently low to permit Q_{R21} to be neglected in this case, a situation which becomes more applicable as the cylinder length to diameter ratio reduces. The full analysis reveals that this cylinder achieves a temperature of 450°C at a discharge power input of 1 W cm^{-2}.

Temperatures derived using this analysis are, of course, for steady state conditions, which will be achieved after different times depending on the power input, specimen size, mass and specific heat. The minimum heating time can be estimated by calculating the thermal capacity $mC_p(\Delta T_1)$, and dividing by the power input. However, in practice the time taken is lengthened by simultaneous heat dissipation. The time taken to reach steady state is longer for lower power inputs and lower ultimate temperatures. For example, when a steel sample of mass 40 gms was subjected to discharge power inputs of 3.5 W, 10.6 W and 94.4 W, the ultimate temperatures of 143°C, 280°C and 640°C were reached in 63, 46 and 18 minutes respectively.

In many plasma assisted processes, the discharge power inputs used ensure that the steady state temperature predicted by this thermodynamic analysis is reached well within the total process time. However, when process times are short it must be borne in mind that the maximum predicted temperatures may not be achieved.

Another practical detail which must be mentioned is the difference between real and idealised samples, particularly the assumption which is implied by this study that the current density is uniform over the sample surface. This is not the case in practice, firstly because of field effects and also because of variations in ion concentrations due to differences in the relative surface area of the edge of the dark space compared to the surface area of the sample. Researchers in Finland have identified this as an important effect, particularly in enhanced plasma nitriding[7,8].

If we consider the situation shown in Fig.3. D minus D' and d minus d' both give twice the cathode fall distance. Ions diffusing across the edge of the cathode dark space are accelerated to the sample, and it is usually assumed that their density is uniform over the sample surface. However, we can see that if N ions per unit area diffuse into the cathode fall, then the ion concentration at the specimen surface will be $\pi d'N/\pi d$ at the thin diameter and $\pi D'N/\pi D$ at the larger diameter. Thus the increase in bombardment occurring at the small diameter compared to the large diameter will be Dd'/dD'. Suppose, for example, d is 0.5 cm, and D is 10.0 cm. If the cathode fall distance is 2.0 cm then d' = 4.5 cm and D' = 14.0 cm. It can thus be seen that the thin section receives an ion concentration over 7 times that received by the larger section. The implication of this for the thermodynamic analysis is that when monitoring a total cathode current to several samples of different size it must be noted that those having a thinner section receive a higher current density than thicker sections, and will thus be heated more. A further implication is that if the discharge could be modified by some means to reduce the cathode dark space distance then heating uniformity would improve. Previously a pressure change would have been considered the only means of achieving this. Recent work now suggests that ionisation enhancement techniques offer a more flexible approach to cathode dark space control[9,8]. Enhancement also offers considerable potential for other aspects of process improvement as discussed in the next section.

FINAL COMMENTS AND CONCLUSIONS

The analysis detailed in this paper provides a basis from which those involved in plasma assisted processes can work on the control of specimen temperatures. However, space has prevented certain aspects being discussed fully. One such is the possibility of forced cooling being applied to samples, for example by direct clamping to a water cooled electrode. This has been considered elsewhere[3]. However, such a cooling arrangement may be difficult to achieve in practice as the contacting face will not be treated and also non-uniformity in heating may result.

Probably the most exciting development in plasma treatment processes (and this includes ion plating and ion nitriding) has been the possibility to utilise lower bias potentials, whilst maintaining the necessary current density by discharge enhancement and support techniques[9]. This provides the means of reducing the power input to samples and hence their ultimate temperature. The author has deposited dense, hard and equiaxed titanium nitride compounds at substrate temperatures below 350°C, when conventional structure theories suggest that temperatures of over 1000°C would be needed. He achieved these results by using a negatively biased thermionic emitter to increase the ionisation in the discharge[10].

This technique ensures that it is not necessary to increase the pressure or specimen bias in order to increase the current density. The important variables are therefore separately controllable and this provides a much better level of control over the discharge power, which this analysis has shown to be the dominant heating source even in ion plating. In that process bias levels of several kV were previously considered necessary, whereas the author has produced coatings having excellent properties at bias levels below 400 V[9]. In ion nitriding also, researchers using a similar technique[8] have produced impressive results at low bias voltage and pressure. In essence then it now appears that the specimen voltage can be used to provide an independent control of temperature, by incorporating ionisation support and utilising the temperature prediction technique presented here.

ACKNOWLEDGMENTS

The author wishes to thank Mr H A Sundquist for many stimulating discussions in this area. Thanks are also due to Dr J Edwards for assistance in formulating the radiation equations, to Mr D G Teer for facilitating and encouraging the work and to Professor J Halling and Professor

W D Morris for their enthusiastic support. The work was carried out as part of an SERC funded programme.

REFERENCES

1. B. A. Movchan and A. V. Demchishin: Phys. Met. Metollogr., 1969, 28, 83-90.

2. J. A. Thornton: Ann. Rev. Mater. Sci., 1977, 7, 239-260.

3. A. Matthews: Vacuum, 1982, 32, 311-317.

4. A. Marciniak and T. Karpinski: Proceedings of the 18th Conf. on Heat Treatment of Materials, Detroit 1980, 334-349.

5. A. Marciniak and T. Karpinski: IPAT 81, Poster Session, Amsterdam 1981.

6. H. Yoshioka: The Seimitsu Kikai, 1981, 46, 9-14.

7. H. A. Sundquist: Private Discussion, Finnish Technical Research Centre (VTT).

8. A. S. Korhonen and E. H. Sirvio: Proceedings or the ICMC 82 Conference, to be published in Thin Sold Films.

9. A. Matthews: PhD Thesis, University of Salford, 1980.

10. A. Matthews and D. G. Teer: Thin Solid Films, 1981, 80, 41-48.

H A SUNDQUIST, E H SIRVIO, and M T KURKINEN

Wear of metal working tools ion plated with titanium nitride

SYNOPSIS

The tribological properties of ion-plated titanium nitride coatings have been studied with a pin on disc, dry sand/rubber wheel and bending under tension friction tests. The test results were compared with the behaviour of these coatings in industrial metal working processes. The coatings proved to increase the life of the cutting tools 2 - 10 times compared with uncoated tools which have the same cutting parameters. The increase in life in the metal forming process studied was even higher.

THE AUTHORS

Mr. Sundquist is at the Technical Research Centre of Finland, Mr. Sirvio at the Department of Mining and Metallurgy at the Helsinki University of Technology and Mr. Kurkinen at Kymi-Kymmene Metal Oy.

INTRODUCTION

The wear of tools in metal cutting and forming industry is one of the main factors in the production costs of machine parts and other refined metal products. Particularly the endurance of cutting tools is very difficult to predict, because so many factors affect the wear of the tools. Some of these are, for example, cutting parameters used, the stiffness of the work machinery, the selection of cutting tool materials, and a large selection of work materials. In metal forming the role of lubricants used and their boundary lubrication properties have a pronounced effect on tool wear.

In metal cutting as well as in wood cutting work materials contain particles with a hardness much higher than that of the workpiece as a whole. Many of these secondary particles are harder than the matrix in high speed tools and retain their hardness to higher temperatures. Such particles in the work material may cause abrasion, especially on steel tools, under conditions of seizure as well as under sliding conditions. In most papers on cutting tool wear the majority of flank wear, in particular, is attributed to abrasion /1/.

In metal forming, fluid film lubrication is not always sufficient to prevent adhesion between the tool and the workpiece. If adhesion occurs, the product is scratched and the deformation process becomes unstable. The wear particles formed are strain- or work-hardened and may cause abrasive wear of the tool.

In abrasive wear the best way to increase the life of a tool is to increase its hardness. When the hardness is increased to exceed the hardness of abrasive particles, the wear of the surface diminishes steeply /2/. However, the ductility of materials usually decreases with increasing hardness affecting the wear resistance. The abrasive wear of hard materials such as ceramics tends to increase with a decrease in ductility and consequent increase in hardness /3/.

If a steel tool is coated with a thin, hard and well adhered ceramic coating, the ductility of the surface can be retained in spite of high hardness. Also the occurrence of steel to steel contact and galling in metal forming can be avoided with these coatings. Another advantage in the use of hard coatings on ductile steel tools can be found in the applications with interrupted cutting process. The brittle fracture of the tool can be avoided at the beginning of the cutting process as for example in grooving.

When thin coatings are used, their minimum thickness has to be determined by the surface roughness of the tool. The coating thickness has to be a few times higher than the R_a value of the surface roughness, otherwise the abrasive wear starts on the tips of the asperities, revealing the base material and wear proceeds quickly /4/.

Titanium nitride coatings have recently been used successfully as hard ceramic coatings on metal cutting and forming tools. The development of plasma enhanced deposition processes has made it possible to deposit titanium nitride on tool steels using low deposition temperatures /5/.

Titanium nitride coated high speed steel twist drills show improved wear resistance especially

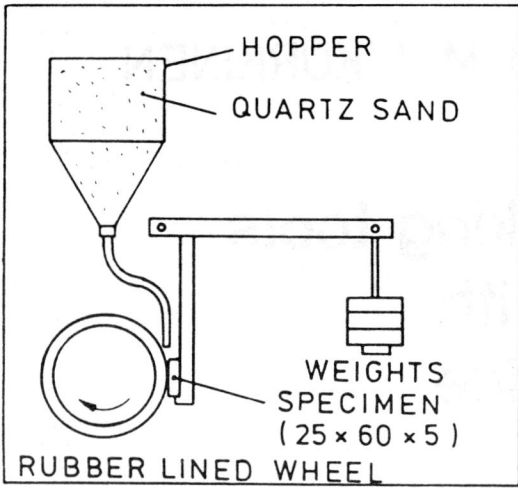

1   Schematic illustration of the dry sand/ rubber wheel test.

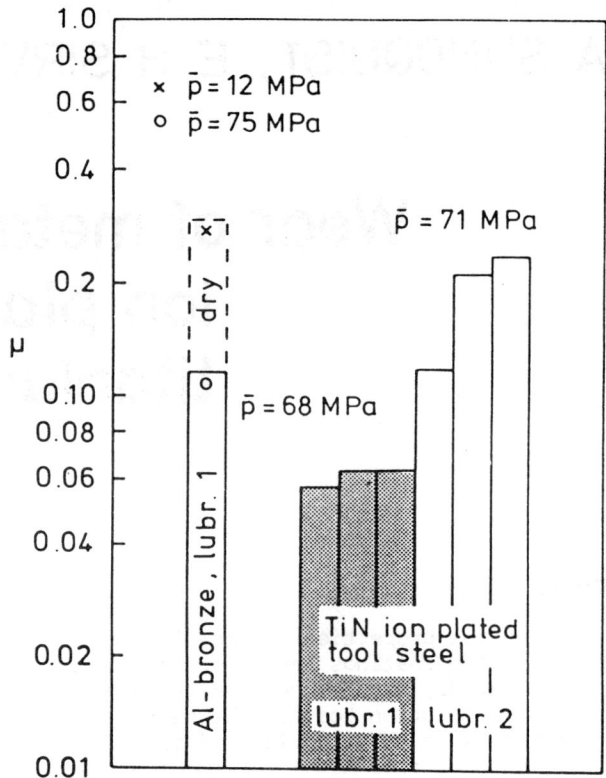

2   Principle of the bending under tension friction test.

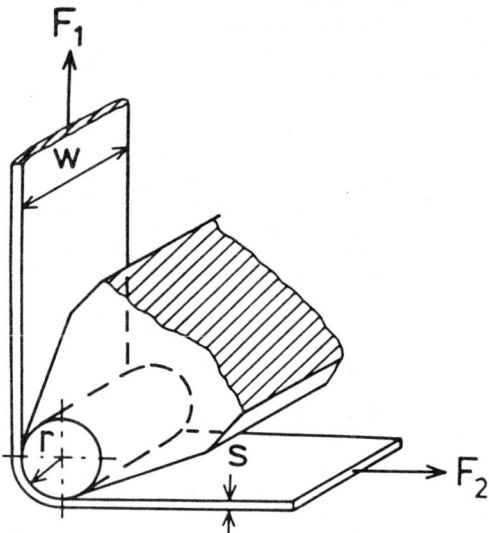

3   Coefficients of friction obtained in the bending under tension friction test.

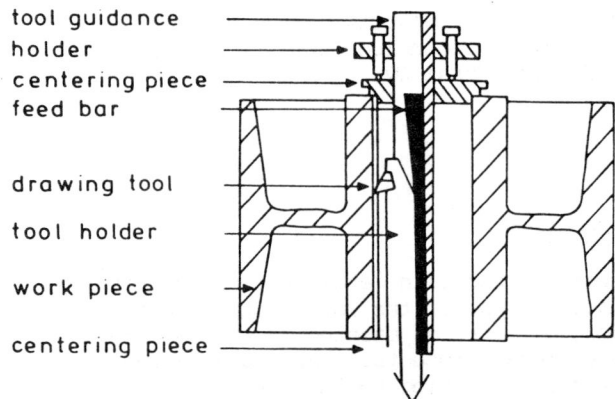

4   Schematic illustration of the key way drawing operation.

when hard abrasive steel is used as a work piece /6, 7/. Flank wear of hob and pinion cutters has been reduced to less than one fifth by using titanium nitride coating on high speed tools /8/. In interrupted cutting tests, cemented carbide tools with titanium nitride coating deposited by PVD techniques have been shown to last much longer than those coated by CVD techniques. Prolonged life has been explained by the higher fracture toughness of these coatings compared with coatings prepared by CVD /9/. These examples show some of the advantages gained by using ion-plated titanium nitride coatings in metal working industry.

In this work the tribological properties of these coatings, and the wear of ion-plated metal working tools, has been studied with laboratory tests and tests in practical metal working operations.

The wear rate and coefficient of friction of these coatings in dry rubbing wear against mild steel have been studied in a conventional pin on disc test. The abrasive wear resistance of the coatings has been studied with a dry sand/ rubber wheel test, and the frictional properties of the coating against steel sheet in plastic condition has been studied in a bending under tension type strip drawing test. In practical applications the wear of coated tools has been compared with that of uncoated tools. Applications studied were key way drawing, gear generating shaping, gear hobbing and sheet bending.

All the coatings tested were deposited by a reactive ion-plating process using a coating thickness of about 4 μm /10/.

LABORATORY TESTS

The following basic properties were studied in laboratory tests: dry friction against steel, adhesive and abrasive wear resistance and friction against plastically deforming steel. The tests used were pin on disc, dry sand/rubber wheel and strip drawing tests.

Pin on disc test

Coated, spherical tipped (diameter 10 mm) pins were loaded against mild steel plate. Bulk material of the pins was X 40 CrMoV51 in hardened and tempered (350°C) condition, hardness 50-55 HR$_C$. Tempering was done during the coating operation, and the reference specimens (uncoated) were made by grinding off the coating. Pins were loaded with dead weights and the loads used were between 0.5 and 10 N. The plate was rotating with a rotational speed of 76 rpm giving a sliding speed of 0.2 m/s on a wear track with a diameter of 50 mm. The overall wear of the pin and the plate were measured by an inductive approach gauge and the coefficient of friction with a load cell.

In these conditions the friction coefficient of uncoated pins was with all loads about 0.7. With a load of 0.5 N the coefficient of friction of the coated pins remained below 0.2 throughout the test, i.e. the sliding distance 1000 m and the approach measured was less than 10 μm. With a load of 1 N the coating wore through after a sliding distance of 150 m. This was detected by an adhesive wear scar in the middle of the wear track and a slow increase in friction. The coefficient of friction increased to 0.4 and remained there to the end of the test. With higher loads the coating wore through at the beginning of the test, and the coefficient of friction increased. With a load of 10 N the coefficient of friction rose to 0.7, i.e., the value of uncoated pins.

In the tests where coated pins gave smaller coefficients of friction than uncoated pins, the approach rate was also smaller. This suggests that a TiN coating on the edges of a wear scar supports part of the load, decreasing the adhesive wear rate and friction coefficients. Actually, this is the situation prevailing on the flank face of a cutting tool.

Dry sand/rubber wheel tests

Figure 1 shows a schematic illustration of the test equipment used. The abrasive, quartz sand, is introduced between the test specimen and a rotating wheel with a chlorobutyl rubber rim with a hardness of Durometer A-60. The test specimen is pressed against the rotating wheel with a force of 130 N while a flow of grit (about 300 g/min) abrades the test surface. The specimens were weighed before and after the test, and the loss in mass was recorded.

The tests were made according to ASTM standard /10/ with some modifications. The differences to standard procedures were as follows (value of corresponding standard test parameter in parentheses): diameter of the rubber wheel 185 mm (228.6 mm); number of revolutions used in each test 420 (100). This gives a sliding distance of 244 m (71.8 m).

The normal force on the specimen was the same as indicated in the standard, i.e. 130 N. The size of quartz sand particles used varied mainly (80%) between sieve sizes 50 and 150, corresponding to diameters from 300 μm to 100 μm.

The test specimens were made of AISI D3 type tool steel X 210CrW12 in hardened and tempered condition. Three different surface roughnesses were machined to evaluate the effect of surface roughness on wear of hard coating. The wear of coated specimens compared to the wear of uncoated steel varied from 4 % to 23 % as a function of the initial surface roughness of coated specimens. With a R$_a$ value of 0.06 μm the relative wear was 4%; with 1.1 μm it was 6%; and with 9.5 μm it was 23%.

Strip drawing test

To evaluate the friction properties of TiN coatings in metal forming processes a bending under tension type test was used, figure 2. The tests were carried out in a frictometer developed by J. Kumpulainen, A. Ranta-Eskola and M. Sulonen /11/. In the test the strip is drawn over a cylindrical bead by force F$_2$, figure 2. The blank tension is adjusted to prevent sliding until plastic deformation occurs. The difference in tension is equal to the sum of the friction force and the force needed to bend and unbend the strip. When the sum of bending and unbending forces has been measured with a freely running shaft instead of a draw bead /12/ the friction coefficient and the contact pressure can be calculated.

The tests were made with nominal contact pressures of about 70 MPa using the lubricants Rocol RTD (lubricant 1 in figure 4) and Stanco Base (lubricant 2 in figure 4).

The behaviour of coated tools was compared with that of aluminum bronze tools known to have good friction and antigalling properties.

From the results given in figure 3, it can be seen that titanium nitride coated tools compare well with aluminium bronze. The coefficient of friction against plastically deforming stainless steel sheet was very low, about 0.06 with Rocol STD lubricant.

WEAR TEST ON TOOLS

To evaluate the advantages of titanium nitride coating on tools the wear of coated tools was compared with that of uncoated tools in practical metal working operations. These tests were made with coated and uncoated tools in production lines containing cutting and forming operations.

Key way drawing

Key way drawing in gears made of ductile cast iron with austenitic-bainitic microstructure /13/ creates problems because of the wear of

5   Gear generating shaping process.

6   Dimensions of the bending tool and workpiece
    used.

Table 1   Cutting parameters and wear of tools for key way drawing.

Tool			Work piece hardness, HB	Key way		cutting		No. of key ways drawn	W_f, mm
	γ	α		b, mm	l, mm	v, m/min	s, mm		
uncoated	4°	6°	341	36	126	6.5	0.09	1	>0.5
coated	4°	6°	341	36	126	6.5	0.09	8	0.2
uncoated	3°	6°	348	28	100	6.5	0.09	1	0.2-0.3
coated	3°	6°	348	28	100	6.5	0.09	10	0.1-0.2

where γ – rake angle   v – drawing speed

         α – clearance angle   s – feed/stroke

         b – width of a key way, W_f – length of flank wear
             tolerance P9

         l – length of a key way

Table 2   Cutting parameters and number of gears produced without
          resharpening.

Cutter $\varnothing$ 100 mm z = 33	Work piece hardness, HB	Cutting conditions				No. at gears produced	W_f, mm
		V_1, m/min	V_2, m/min	S_w mm	S_r mm		
uncoated	311	9	12	0.5	0.03	4	0.2
coated	311	9	12	0.5	0.03	18	0.2

where V_1 – cutting speed 1st cut S_r – infeed/stroke

         V_2 – cutting speed 2nd cut W_f – flank wear

         S_w – rotary feed

high speed steel tools. In figure 4 there is a schematic illustration of this drawing process.

Coated and uncoated tools made of steel S10-4-3-10 with a hardness of 65 HR$_C$ were used. Table 1 gives the cutting parameters, the number of key ways made without resharpening of the tool, and the flank wear measured.

These key ways had previously been drawn by cemented carbide tipped tools using the cutting parameters shown in table 1. During these tests it was found that high speed steel tools coated with titanium nitride gave about the same endurance if the tool is carefully resharpened. The endurance was almost tenfold compared with uncoated steel tools.

Gear generating shaping

In this operation sintered high speed steel (HSP 23) with a hardness of 67 HR$_C$ was used to produce internal gears by the shaping process illustrated in figure 5. The cutting parameters, the number of gears produced without resharpening, and the flank wear measured are in table 2.

The endurances of coated tools in this generation were, according to table 2, 4.5 times longer than those of uncoated tools. The flank wear of coated tools after three machined gears was already about 0.1 mm. However, then the wear rate decreased, and thus it is obvious that the difference of the endurance between coated and uncoated tools would have been even greater with 0.3 mm chosen as the limiting flank wear.

Gear hobbing

In gear hobbing the accuracy of the completed gear depends on the condition of the tool during the last finishing cut. Thus it is greatly affected by the wear of the hob. Wear resistant coatings may be a good solution to this problem. When using sintered solid steel hobs, ion-plating has proved a suitable coating process and titanium nitride a beneficial coating material.

In this study sintered solid steel hobs (HSP 23) with a protuberance profile were used. In these tools the flank wear is a serious problem, because the profile leads inevitably to a small clearance angle. The profile is, however, needed to produce a root relief for case hardened gears when only the teeth faces will be ground after hardening.

The flank wear of coated and uncoated hobs was studied in a production line producing gears with moduli of 4, 5 and 6. The cutting conditions were kept constant in both cases. In these tests 2 to 3 times more gears were cut between resharpening operations with coated tools than with uncoated tools.

Sheet bending

The sheet can be bent with a tool (illustrated in figure 6) to produce light hinges. The bending groove of the tool has to be ground and polished to ensure good surface finish. The only lubricant used in this process is the oil which remains on the sheet surface after rolling.

Uncoated tools have to be reground and polished after a production of 1000 hinges.

10,000 hinges could be made with a coated tool continuously without any scratching of the product. The process could be continued five times when the adherent wear residues were removed from the tool surface. After the production of 50,000 hinges the coating wore through, and the tool had to be reground, polished and recoated.

In this application the titanium nitride coating proved to be very useful, because during the lifetime of the coating, 45 grinding and polishing operations were made unnecessary. The useful life of the coated tool could thus be considered to be 50 times that of an uncoated tool.

DISCUSSION

Ion-plated titanium nitride coatings on tool steel surface decrease the wear rate of the tool considerably. Laboratory tests showed that the wear rate and the coefficient of friction decreased in adhesive wear (pin on disc test) even though the coating had been worn through. This must be the result of the high load carrying capacity of coating at the edges of the wear scar resembling the situation in flank wear on a cutting tool.

Abrasive wear of tool steel can also be considerably decreased by a TiN coating. In this application, however, the initial surface roughness of the tool has an important role in tool wear. To increase the life of the tool the surface roughness must be less than the coating thickness.

The coating effectively protects plastically deforming steel against high adhesive forces between the tool and the workpiece. In friction test in this study an uncoated tool would have been seriously damaged.

These findings of laboratory tests have been confirmed in practical metal forming and cutting processes. In the cutting processes studied the cutting parameters used were the same as for uncoated tools during conventional production. The endurance of the coated tools was 2 - 10 times that of uncoated tools depending on the process studied.

Further work is still needed to find the optimum cutting parameters for coated tools. For example with an increase in feed or cutting speed the increase in the volume of material cut in one sharpening period compared with that with uncoated tool can probably be increased more than the life of the tool when using the same cutting parameters. An increase in feed also decreases the harmful effects of rake face wear by moving the wear crater further up from the tip of the tool. This may be an important aspect, because the low adhesion of the coating to steel decreases the volume of dead metal on the tip of the tool, bringing the rake face wear crater nearer the tip.

In metal forming operations the beneficial effects of the coating were many times those obtained in cutting operations. This is explained by the good antigalling properties of coated tools compared with those of uncoated tools.

The process of low temperature reactive ion-plating of titanium nitride is thus shown to be of great value to the metal working industry. If the deposition temperature remains below 350°C, the coating of several tool materials is possible. Consequently, TiN coatings can be used also on metal forming tools, where their good tribological properties can readily be used to increase the tool life.

ACKNOWLEDGMENTS

The study reported in this paper formed part of a wider investigation which was supported by the Academy of Finland. The authors wish to thank Mr. D.G. Teer and Dr. A. Matthews for their help in developing the coating process, and Mr. T. Haikola for valuable discussions on the wear testing of the coatings.

REFERENCES

1   E. M. Trent, Treatise in Materials Science and Technology, 13 (1979), 443-489.

2   H. Uetz, J. Föhl, Wear 49 (1978), 253-264.

3   K.-H. Zum Gahr, Metal Progress 116 (1979) 4, 46-52.

4   E. Sirvio, H. Sundquist, Abrasive wear of ion plated titanium nitride coatings on plasma nitrided steel surfaces. Paper presented at ICMC-82, San Diego, 1982.

5   A. Matthews, D. G. Teer, Thin Solid Films 72 (1980), 541-549.

6   R. Buhl, H. K. Pulker, E. Moll, Thin Solid Films, 80 (1981), 265-270.

7   R. R. Nimmaggadda, H. J. Doerr, R. R. Bunshah, Thin Solid Films, 84 (1981), 303-306.

8   A. Kono, Technocrat 14 (1981) 2, 32-38.

9   M. Kodama, R. F. Bunshah, Interrupted cutting tests of cemented carbide tools coated by PVD and CVD techniques. Paper presented at ICMC-82, San Diego, 1982.

10 ASTM Standard, G65-80, 1980.

11 A. Ranta-Eskola, J. Kumpulainen, M. Sulonen, Comparison of strip drawing tests used for measuring surface interactions in press forming. Paper presented at 12th Biennial Congress, IDDRG, Genoa, 1982.

12 M. Sulonen, P. Eskola, J. Kumpulainen, A. Ranta-Eskola, IDDGR Working Group meeting, Japan 1981, Paper IDDRG/WG III/4/81.

13 M. Johansson, AFS Transactions, 73 (1977), 117-122.

G CARTER

Ion implantation and ion beam mixing

SYNOPSIS

Processes of ion implantation and ion beam mixing are becoming increasingly important for the surface property modification of metals. This paper outlines the physical and metallurgical processes which occur during ion penetration and atom recoil production in a solid and defines possible limits to the efficacy of the process. A summary is then given of major applications of the technique in near-surface physical and chemical changes to metals. Some industrial applications are discussed and the techniques are compared to other surface preparation processes.

THE AUTHOR

Professor Carter is in the Department of Electronic and Electrical Engineering, University of Salford, UK.

INTRODUCTION

The interaction of energetic ions with solids has been the subject of quite intensive scientific investigation for over 100 years {1}. Only in the last 25 years or so, however, has the deliberate and controlled injection of ions into solids become of very considerable technological import- ance. Ion injection and extrapment was first employed as a technique of removing atoms from the gas phase in the Ion Pump {2}, a still out- standingly successful application.

Following some 15 years of investigation, false starts, realistic and unrealistic expectations and very considerable commercial resistance in some quarters, the controlled and defined injec- tion or implantation of suitable electrically dopant atoms into semiconductor substrates came to be an accepted industrial manufacturing technique in the early 1970's. Contemporary submicron size solid state devices would be virtually impossible without the use of ion beams.

At about the time of penetration of ion implantation into the semiconductor manufacturing market, a number of investigators began to explore the possibilities of employing similar techniques to modify the properties of metals and significant progress has been made upon understanding the physical and metallurgical processes that occur. The following section will review the current status of this knowledge. However very little industrial uptake of this technique has yet occurred but some success stories will be reported in the third section of this review.

The process of ion implantation is a violent one and situations far from thermodynamic equili- brium result. Thus collisions of the energetic projectiles with substrate atoms can lead to the dynamic recoil of the latter creating lattice defects, both point and extended and leading to a spatial redistribution or mixing of atoms. This mixing process has become the topic of quite substantial current investigation and again in the following section we review present understanding of the process and in the third section outline interesting results obtained.

In the final section some comparisons with other surface treatment processes will be made.

PHYSICAL AND METALLURGICAL ASPECTS OF ION IMPLANTATION AND MIXING

When an ion, of energy greater than a few keV, strikes a solid surface it has a probability, approaching unity, of entering the surface {3} and slowing down to rest, following a sequence of collisions with atoms and electrons of the substrate. The nature of collision and energy loss processes is rather well understood theoreti- cally {4,5} and confirmed experimentally {6}, particularly for most ion species in the energy range 10 keV-500 keV. Because the sequence of projectile-substrate atoms is statistically distributed, the final resting places of a large number of incident projectiles will also be statistically distributed spatially and particu- larly in depth into the substrate. Fig. 1 displays, schematically, such a range distribution.

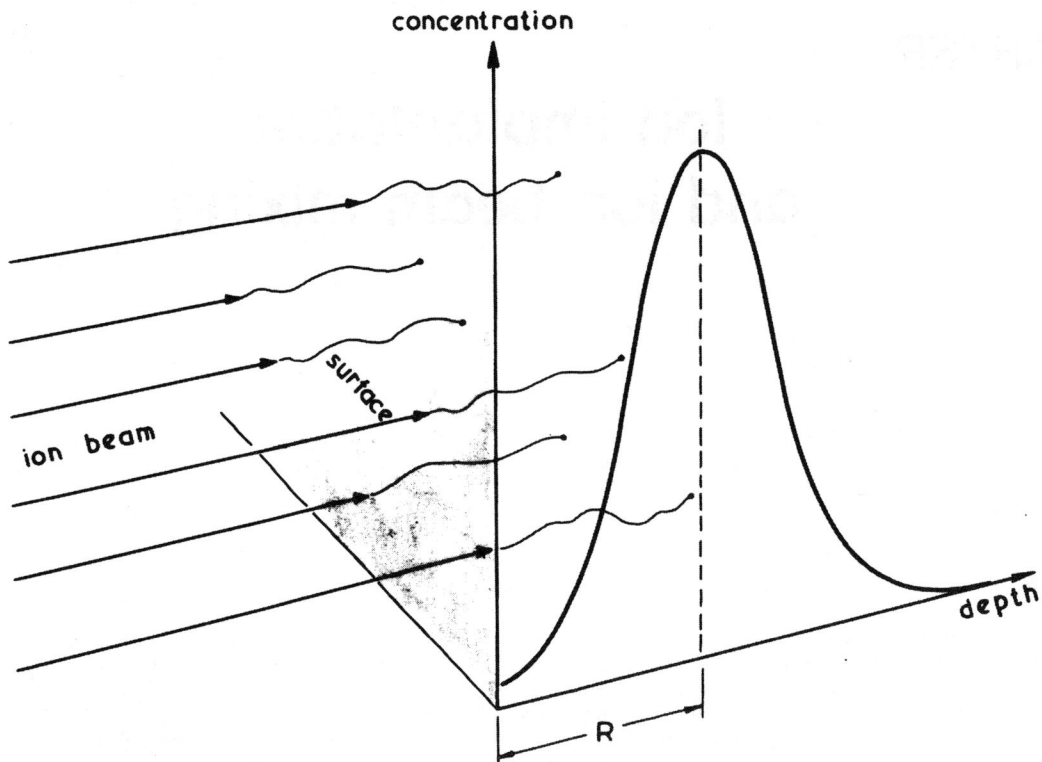

1. Schematic illustration of the depth concen-
    tration profile of a low ion fluence implan-
    ted into a solid. The range R is indicated.

Because collision processes are rather well understood the form of such range distributions can be quite accurately predicted and extensive tabulations exist for the moments (i.e. mean, variance, skewness, etc) of such distributions {7,8}. As a very rough guide the mean range (and the standard deviation) are of order 0.1-5.0 nm keV^{-1} of ion energy but precise values vary with ion and substrate species (and orientation). The behaviour of these moments, in the energy range cited, with incident projectile energy, E_0, is like $E_0^{2/3}$ to E_0. This immediately points up a potential problem of ion implantation that with accelerators of reasonable energy capability (and cost, safety, etc), the depth to which foreign atoms may be injected is usually only of the order of µm or less. In semiconductor applications this is acceptable but for metals in hostile environments (e.g. wear, corrosion situations) there may be initial doubts about whether such thin layers which may be potentially influenced by incorporation of foreign atoms, are of technical significance. It will be seen later, that, in some circumstances, the depths influenced by implantation can be much greater than the penetration range so that implantation cannot be dismissed a priori as ineffectual.

As a projectile slows down in the substrate it will lose typically about 70% of its energy in a succession of collisions with matrix atoms. If in collision an energy greater than a dynamic

threshold energy, E_d, of about 20-50 eV is trans-ferred to a matrix atom this latter can recoil from its lattice position as a dynamic interstit-ial and leave a residual vacancy. If energy transfers are large the dynamic interstitial or primary recoil can initiate a cascade of higher generations of recoils so that a rather large local density of vacancy-interstitial pairs is generated. A rough estimate of this defect density is $\dfrac{5\,E_0}{N E_d R^3}$,

where N is the atomic density and R is the ion mean range, which for 100 keV ions leads to a defect fraction of 10^{-4}-10^{-2} depending upon ion and target species. This, of course, is an enor-mous defect level compared to those induced thermodynamically and although defects will be lost by migration, recombination and agglomeration it would be no surprise if even a fraction of this defect concentration were to lead to changes in materials properties. Under many circumstances these defects will be isolated point defects but for ~100 keV heavy atomic and molecular ions, the recoil generation frequency along the projectile track is high and vacancies are created in close proximity with interstitials being ejected trans-verse and parallel to the track. Under these conditions it appears {9} that the vacancy rich core of the collision cascade or spike can coll-apse to a vacancy dislocation loop. Under less severe local defect production densities it is often observed {9} that the interstitials generated can migrate to form interstitial loops.

Under all circumstances, however, it is expected that the depth distribution of defect creation will be similar to, but shifted towards the surface with respect to the projectile range distribution {5}. However, the centroid of the stopping positions of the dynamic recoils will be deeper in the solid than that of the vacancy population {10,11} and thus defect gradients are established. If these defects are thermally mobile then they may cause transport of impurity solute atoms in spatial directions depending upon the impurity-defect binding type and energy {12, 10}. This process is known as radiation enhanced diffusion.

In addition, however, since recoils are kinematically transported (over rather short distances on average since recoil mean energies are low) in each collision sequence, then after many ion impacts each matrix atom within the stopping range of the ions will have been relocated many times. This is equivalent to a quasi-diffusional or mixing process and is known as physical atomic mixing. One component, known as recoil mixing, results from direct projectile-substrate or impurity atom collisions, another arises from collisions between recoiling substrate atoms and other substrate or impurity atoms and is known as cascade mixing. Diffusional approximation {13-16} and integral equation transport approximation {17-20} models have been developed to support experimental studies in this area.

In view of the above 'enhanced atomic relocation' processes it is hardly surprising that bilayer and multilayer thin metal film structures have been successfully intermixed, that initial solute profiles should be modified, that precipitates should be either destroyed or formed, and that, during continuing high fluence implantation, the implant accumulation depth profile should be perturbed. The generation of high local defect densities and the possibility of incorporation of large concentrations of implanted impurity might also be expected to lead to phase change phenomena. Examples of all these processes will be given in the next section.

Finally a further aspect of the radiation damage or recoil atom generation phenomenon should be mentioned, that of surface atomic erosion or sputtering. If, as a result of the sequences of projectile or recoil atom collisions, atoms at and near the solid surface receive sufficient directed energy to overcome surface binding forces then they will be ejected. The experimental and theoretical literature in this topic is extensive {21} but for present purposes one parameter is crucial, the sputtering yield, Y, or atoms ejected per incident ion.

For many ion-substrate systems Y increases from zero at an incident energy of a few tens of eV, reaches a maximum, often greater than unity at an energy in the 10-100 keV range and then declines. Very simple arguments {22} indicate that the maximum impurity concentration that can be achieved during simultaneous implant collection and sputtering is of order Y^{-1}. Thus if Y is large (e.g. 10) then only small fractional impurity concentrations may be achieved, whereas smaller values of Y lead to higher concentrations. It is for this reason that relatively energetic ($\gtrsim$50 keV) ions are preferred for implantation into metals {23} or very low energy ions {24,25,26} ($\lesssim$100 eV) are employed so that overlayer coatings may be achieved by 'implantation' of a growing film.

The sputtering of alloys is a much more complex problem than elemental solids {27} and since a heavily implanted metal may be thought of as an 'alloy system', the simple relationship between concentration and sputtering yield indicated above can only be taken as a rough guide. Nevertheless the much higher concentrations expected to be needed to substantially modify near surface properties of metals compared to the low concentration requirements of semiconductor applications may, again, give some doubt about the potential efficacy of the technique in metals processing. As will be shown shortly however, significant improvements may be achieved in various near surface properties of metals at the $\lesssim$10% impurity concentration level and the technique should not be immediately dismissed.

It is informative, also, to note {22} that the ion fluence required to achieve the maximum concentration is of order $\frac{NR}{Y}$, which generally leads to values in the range 10^{16}-10^{17} cm^{-2}. Such fluence levels may only be achieved in acceptable technical processing times by using rather large ion flux densities. Thus an ion current density of 1 mA cm^{-2} will produce a saturation concentration in times of the order 1-10 sec. Such high currents carry the penalty of high power loading (e.g. 100 W cm^{-2} for 100 keV ions) at the solid surface and potentially unacceptable temperature rise without careful cooling.

SOME APPLICATIONS OF ION IMPLANTATION AND ION BEAM MIXING

It would be unjust to a wide range of authors to attempt a complete catalogue of the various interesting and successful applications of these techniques to metals processing. Several recent, comprehensive reviews of various applications are published and the reader is referred to these {23,28-33}. Here we will present an overview of the areas in which investigations are proceeding and highlight some of the more promising activities.

ION IMPLANTATION

Since ion implantation leads, essentially, to rather near surface changes in atomic composition and defect concentration, it is the near surface properties of solids which may be considered amenable to potential modification. Two types of property modification have been explored, the first concerned with mainly physical tribological processes such as friction, wear, surface hardness and fatigue, the second with mainly chemical (or electrochemical) processes such as aqueous and high temperature corrosion and catalysis. In this latter area one may also include phase change phenomena such as the conversion of certain metals from crystallinity to amorphousness via the introduction of large concentrations of chemical dopants.

Hartley {32} and Dearnaley {28} have summarised studies of modification of frictional properties of steels, permalloy and cemented tungsten carbide. In steels both enhancements and reductions in coefficient of friction were achieved with different implant species, observations explained by Hartley {32} in terms of the yield stress of deformed junctions at implanted-non implanted metal during sliding. In the more recent studies of Co-cemented tungsten carbide by Dearnaley and Charter {34}, 4 x 10^{17} cm^{-2}, 90 keV N$^+$ implantation resulted in a factor of almost 2 reduction in coefficient of friction when sliding against steel, a factor maintained for long periods.

Tentative explanations of the effect of N^+ implantation were in the production of a low friction Beilby layer which was inhibited from adhesive wear removal by either strengthening the cobalt and increasing hardness or improving adhesion between the cobalt and tungsten carbide grains.

Intimately associated with friction is the problem of wear and probably the most extensive investigations, recently reviewed by Hirvonen [33] and Dearnaley [28,29] have been conducted in this important technological area. The early work by Hartley [32] established that the wear of lubricated steel was much reduced by $\geq 10^{17}$ N^+ or C^+ ions cm^{-2} at energies ≥ 50 keV and that, particularly in the case of N^+, 10-20% of the nitrogen was still located in the solid ahead of the wear front after erosion of 1 to 5 µm (compare this with the much smaller implant range). Dearnaley and Hartley [35] suggested that N, implanted interstitially, could segregate to dislocations created during wear and then undergo solute drag into the metal as the dislocations propagated inwards. More recently, Dearnaley [36] has also invoked inwards pipe diffusion of N and C along dislocation cores to explain the deep migration and suggested that these dislocations are pinned by segregated N to increase hardness and reduce wear. Dearnaley [28] has also noted studies with other implants into steels which may assist in implanted N trapping and suggests that situations favouring nitride precipitation are to be avoided since these diminish the supply of N to the dislocations. Similar mechanisms were also suggested for improvements in wear resistance of Cr films and the lack of improvement in only N^+ implanted Al and Ti.

Recent studies by the author's colleagues at Salford and UMIST on wear and coefficient of friction improvements in ion implanted Cu have been summarised by Grant and Colligon [37]. It was observed that B^+ implantation gave extended life improvements in both properties whilst N^+ and B^+ implantation resulted in short term benefits.

In the applications area Dearnaley and Hartley [35] have reported four-fold increases in lifetime of N^+ implanted chromium-carbon steel tool punches and cutting knives; twelve-fold lifetime increase of cobalt-cemented tungsten carbide slitting knives; ten-fold increase in steel press tool operation [36] and similar increases in successful operation of a wide range of tools in plastics injection moulding [36]. Less dramatic effects, and even null improvements, have been observed in implanted high speed twist drills [35], in tools for steel cutting [36] and in cobalt-cemented tungsten carbide lathe tools. In all these latter processes higher temperatures and/or more aggressive tool indentation are involved.

In the area of hardness studies Hirvonen [33] has reviewed recent work and shown that N^+ implantation into steel and Fe and B implantation into Be increases microhardness (in the latter case by up to a factor of six following annealing). Other observations of increases in hardness of Ne^+ implanted Fe suggest that defect structures may play a role as well as impurity incorporation.

Fatigue studies have also been summarised [28,29,33] and again the major implant species have been N^+ and C^+ and substrates of steel, Ti and Ti/Al/V alloys have been examined. Thus the rolling contact fatigue life of steel ball bearings was improved by about a factor of two by nitrogen implantation [29], Ti alloy fatigue by a factor of three by N^+ implantation and consider-

ably more by C^+ implantation [29]. In this Ti alloy relatively low dose Ba^+ implantation [38] has been reported to increase endurance by a factor of ten due, it is postulated, to formation of $BaTiO_3$ precipitates which both pin dislocations and reduce oxidation and associated fretting fatigue by blocking inwards oxygen diffusion paths. Some studies with Cu substrates have also been summarised [28,29] where Al^+ and Cr^+ ions improve fatigue life but the effect of B^+ implantation is uncertain.

In the fields associated with chemical attack of surfaces reviews have again been given by Dearnaley [28,29] and Hirvonen [33]. In the corrosion area studies of high temperature oxidation, aqueous corrosion and pitting corrosion are preeminent. Most implants in such cases are metallic species. Dearnaley [30,36] has concluded that implantation is most beneficial in inhibiting thermal oxidation of alloys when the implanted impurities segregate either at the metal-oxide interface and impede cation out diffusion or (as in Ti) at dislocations where they block inwards migration of oxygen. Several ion species (at the 0.5% level) such as Sr^+, Eu^+ and Lu^+ are effective in Ti whilst Y and rare earth species show improvements in stainless steel, Ni and Fe-Cr alloy [39, 40]. Self ion bombardment of Cr and Ni and Al^+, Ti^+ and Cr^+ implants into Cu have been reported [41] to be beneficial. Dearnaley [30] has also reviewed oxidation inhibition effects in implanted metals which probably result from (1) changes in oxide plasticity (Zr), (2) formation of a protective barrier oxide (Al, Al/Cu alloys and Fe-Cr alloy) and modifications in (3) the electronic conductivity of oxide films and (4) processes associated with catalytic behaviour.

Detailed studies of the high temperature oxidation behaviour of Ni^+, Cr^+ and Li^+ implanted Ni have recently been reported by the UMIST/Salford group [41,42]. Self ion implantation exerts both short and long term effects but Cr^+ and Li^+ implantation suppresses initial NiO scale formation by production of LiO_2 or $NiCr_2O_4$ nodules. At later oxidation stages it was concluded that the Li and Cr are incorporated in the equilibrium NiO scale and inhibit Ni^{2+} migration. As a result, steady state oxidation of Ni is enhanced by Cr^+ implantation but retarded by Li^+ implantation.

Pitting corrosion has been reported to be eliminated by Cr^+ implantation into a ball bearing steel [41], whilst Grant and Colligon [37] have reported substantial reductions in tarnishing of Cu exposed to atmospheres containing H_2S following different species implantation of which Cr^+ was observed to be most beneficial.

Rather detailed studies of aqueous corrosion inhibition have been undertaken by the present author's colleagues at Salford and UMIST who showed [44] initially, as did Sartwell et al [45] subsequently, that implantation generated alloys of Cr^+ (and Ni^+) into Fe behaved in a similar manner with respect to corrosion passivity as conventionally prepared alloys. However, Ta^+ implantation into Fe leads to significantly improved passivation [46] and Ashworth et al [47] suggest that it is with type of implant which produces target composites generally unachievable via conventional metallurgical techniques which may lead to to interesting and novel improvements in aqueous corrosion resistance. These authors have also produced [49] an extensive survey of effects of implantation in a wide range of substrates and discussed physical mechanisms in detail.

In the associated area of catalytic and electrocatalytic behaviour of implanted metals relatively little work has been reported but a recent summary is given by Wolff {48} showed that Pt, Au implantation into Fe led to strong increases in the hydrogen redox reaction whilst Pb implantation gave a negative effect. On the other hand Ag implantation into Ni {49} has been found to reduce the oxygen overpotential of Ni, important in the oxygen redox reaction.

Phase change phenomena resulting from implantation have been reviewed by Grant {31}, Kaufman and Buene {50}, Picraux {51}, Borders {52}, Poate and Cullis {53} and Sood {54}. General results of such studies are that at low ion fluences the incorporation of impurity atoms on to specific lattice sites is usually (with some exceptions such as Li in Be) similar to that obtained by thermodynamic means (e.g. the Miedema plot is obeyed). At higher fluences, however, where impurity concentrations become large, interesting effects such as amorphisation occur in B {55}, P, As, Dy and Bi implanted Ni {31}. The systems found to exhibit this amorphisation phenomenon are frequently {31} those which can be achieved by rapid quench techniques but this should not necessarily be taken to indicate that the implantation is equivalent to rapid local thermal excursions.

ION BEAM INDUCED ATOMIC RELOCATION (MIXING)

Some of the more fundamental investigations of radiation assisted redistribution of solute atoms in dilute metal alloys, influenced by defect associated transport and solute segregation processes and the partitioning of solute atoms between precipitates and matrix influenced by recoil mixing and segregation have been reviewed by Marwick {56}. This author has given detailed analytical and quantitative consideration to the atomistics of recoil generation and defect and solute flow processes and found good agreement with experimental studies.

Atomic mixing studies in what eventually become concentrated 'alloys' have been conducted and reviewed by the Caltech group {57,58} employing bimetallic thin film and multilayer thin films through which high energy inert gas ion beams were propelled. Systems investigated include Ni, Pd and Pt with Al; Cu, Ni and Co with Ag; Ni, Co, Fe and V with Au and Cu-Co with a concomitantly wide range of results. Examples of single phase f.c.c. solid solutions were found with Ag-Cu and Au-Co but these transformed to two phase mixtures after annealing. Simultaneous generation of both f.c.c. and b.c.c. structures were found for specific Au/Fe and Au/V compositions whilst irradiation at low temperatures generated amorphous structures in both Au/Co and Au/V mixtures.

Dearnaley {28,29} has also discussed atomic relocation studies with particular reference to defect enhanced inwards migration of thin film metals into substrates with the projectiles injected through the films to induce defect formation in the substrate. Dearnaley {28,29} describes such processes as 'bombardment diffused' and has reported on the introduction of Co and Ir into Au by irradiation with 400 keV N^+ ions. He also discusses {28} bombardment diffused Sn in Ti and indicates that this leads to significant improvements in friction and wear resistance and bombardment diffused Si in Fe (with Ar^+ bombardment) and reports that subsequent oxidation rates of the treated Fe at 600°C were reduced by a factor of 45.

At Salford, Colligon, Hill and colleagues {59, 60} have adopted a somewhat different approach to mixing studies. Whereas in the studies described above the film thicknesses are reduced by both intermixing and sputtering, these investigators have maintained a constant film thickness by continuous deposition from a second sputtered atom source. This technique is reported {60} to have led to increased hardness of Ni mixed into Fe and recent studies have indicated excellent adhesion and electrochemical properties of Au mixed into glass.

In the area of high fluence implantation where mixing is expected to modify implantation profiles no systematic work has been performed with metal-metal systems. Again at Salford however, we {61} have recently studied high fluence implantation in Ge implanted Si and measured the modification in depth profile with increasing fluence. The results both indicate the presence of a redistribution process and correspond reasonably well with transport calculations.

Little work has been undertaken on the accumulation of surface coatings by low energy ion implantation but a recent study by Thomas et al {26} is significant in that rather good quality crystals of Ag have been grown on (111) Si substrates using 25-100 eV Ag^+ ion irradiation at room temperature. Apparently the collisional processes induce crystallographic growth at temperatures much less than required in conventional evaporation-condensation.

COMPARISON WITH OTHER SURFACE MODIFICATION PROCESSES

It is not intended here to present a comparison of results obtained by different processing techniques but to point out the similarities and differences in physical processes in different techniques. Firstly ion implantation acts within the solid and apart from minor swelling associated with atomic incorporation (perhaps of the order 10-100 atomic layers) little dimensional change occurs in contrast to plasma deposition and ion plating processes where surfaces are coated. Despite the apparently limited expected depth of action many examples are known where implantation and mixing effects operate rather deeply and effectively into metals as a result of diffusion processes.

Secondly ion implantation and mixing are hard vacuum processes and thus allow, without extraordinary difficulty, the elimination of impurities, more difficult to control continuously in coating techniques.

Thirdly the action of collisional processes in generating defects, causing intermixing and homogenisation are probably not dissimilar in nature in, on the one hand implantation and ion beam mixing and on the other hand, plasma and ion coating processes. The magnitudes of the processes differ however since the implantation/mixing processes generally employ rather high energy (but somewhat low flux) projectiles whereas coating processes employ lower energy but higher flux projectiles.

A combination of these processes is now under investigation by the author using implanted high energy projectiles of the same species as congruently evaporated film materials to study atomic mixing phenomena and film growth under controlled conditions.

REFERENCES

1. G Carter and J S Colligon. "Ion Bombardment of Solids". (Heinemann Educ. Books, London) 1968.

2. R A Douglas, J Zabritski and R G Herb. Rev Sci Instrum 36, 1 (1965).

3. G Carter, D G Armour, S E Donnelly, D C Ingram and R P Webb. Rad Effects 50, 97 (1980).

4. J Lindhard, M Scharff and H E Schiøtt. Mat Fys Medd Dan Vid Selsk 33 (No 14) (1963).

5. K B Winterbon, P Sigmund and J B Sanders. Mat Fys Medd Dan Vid Selsk 37 (No 14) (1970).

6. S Kalbitzer and H Oetzmann. Rad Effects 47 57 (1980).

7. K B Winterbon "Ion Implantation Range and Energy Deposition Distributions" Vol 2 (Plenum Press, New York) (1975).

8. V Littmark and J E Ziegler. Phys Rev A 23 64 (1981).

9. D A Thompson. Rad Effects 56, 105 (1981).

10. P C Piller and A D Marwick. J Nucl Mat 71, 309 (1978).

11. L A Christel, J F Gibbons and S Mylroie. J Appl Phys 51 6176 (1980).

12. R E Howard and A A Lidiard. Phil Mag 11, 1179 (1965).

13. H H Andersen. Appl Phys 18, 131 (1979).

14. G Carter, D G Armour, D C Ingram, R P Webb, and R Newcombe. Rad Effects Letters 43, 233 (1979).

15. S Matteson, B M Paine and M-A Nicolet. Nucl Instrum & Meth 182/183, 53 (1981).

16. R Collins. Rad Effects Letters (To be published 1982).

17. W D Hofer and V Littmark. Phys Letters 71A, 457 (1979).

18. V Littmark and W D Hofer. Nucl Instrum and Meth 168, 329 (1980).

19. P Sigmund and A Gras-Marti. Nucl Instrum & Meth 168, 339 (1980).

20. P Sigmund and A Gras-Marti. Nucl Instrum & Meth 182/183, 25 (1981).

21. Topics in Applied Physics Sputtering by Particles, Vol 1 (Ed. R Behrisch, Springer-Verlag, Heidelberg) 1981.

22. G Carter, J S Colligon and J H Leck. Proc Phys Soc 79, 299 (1962).

23. G Dearnaley. Rad Effects (To be published 1982).

24. B Probyn. J Phys D 1, 457 (1968).

25. J Amaro, P Bryce and R P W Lawson. J Vac Sci Technol 13, 591 (1976).

26. G E Thomas, L J Beckers, J J Vrakking and B R De Koning. J Cryst Growth 56, 557 (1982).

27. H H Andersen. Proc Symp Phenomena in Ionized Gases, Dubrovnik 1980. (Ed. B Cobic, Boris Kidric Inst Nucl Sci Beograd) (1981).
See also Z L Liau and J W Mayer in "Treatise on Materials Science and Technology", Vol 18, Ion Implantation (Ed. J K Hirvonen, Academic Press, New York) p17 (1980).

28. G Dearnaley. Rad Effects (To be published 1982).

29. G Dearnaley and P D Goode. Nucl Instrum & Meth 189, 117 (1981).

30. G Dearnaley. Nucl Instrum & Method 182/183 899 (1981).

31. W A Grant. Nucl Instrum & Method 182/183, 809 (1981).

32. N E W Hartley. Thin Solid Films 64, 177 (1979).

33. J K Hirvonen. J Vac Sci Technol 15, 1662 (1978).

34. G Dearnaley & S J B Charter. Proc Int Conf on New Frontiers in Tool Material for Metal Cutting and Forming, London, 1981. (Engineers Digest, 120 Wigmore St, London).

35. G Dearnaley & N E W Hartley. Proc 4th Conf on Scientific and Indust Applns of Small Accelerators, 1976. (IEEE, New York) p20.

36. G Dearnaley. Proc Conf on Ion Implantation Metallurgy, Boston 1979. (Eds. J K Hirvonen and C M Preece, AIME, Warrendale, Pa), p1 (1980).

37. W A Grant and J S Colligon. Vacuum (To be published, 1982).

38. G Syers and G Dearnaley (Unpublished).

39. D B Peplow and G Dearnaley. Corrosion Science. (To be published).

40. M Slater, W A Grant and G Carter. To be published in Nucl Instrum & Meth.

41. Y F Wang, C R Clayton, G K Hubler, W H Lucke and J K Hirvonen. Thin Solid Films 63, 11 (1979).

42. Zhou Peide, F H Stott, R P M Procter and W A Grant. Oxidation of Metals 16, 409 (1981).

43. F H Stott, Zhou Peide, W A Grant and R P M Procter. Corrosion Science 22, 305 (1982).

44. V Ashworth, W A Grant, R P M Procter and T C Wellington. Corrosion Science 16, 393 (1976).

45. B D Sartwell, A B Campbell and P B Needham in "Ion Implantation in Semiconductors". (Ed. F Chernow, Plenum Press, New York) p201 (1977).

46. V Ashworth, D Baxter, W A Grant and R P M Procter. Corrosion Science 17, 947 (1977).

47. V Ashworth, R P M Procter and W A Grant in "Treatise on Materials Science and Technology", Vol 18, Ion Implantation (Ed. J K Hirvonen, Academic Press, New York) p175 (1980).

48. G K Wolf. Nucl Instrum & Meth 182/183, 875 (1981).

49. V Akano, J A Davies, W W Smeltzer, I S Tasklykov and D A Thompson. Nucl Instrum & Meth 182/183, 985 (1981).

50. E N Kaufman and L Buene. Nucl Instrum & Meth 182/183, 327 (1981).

51. S T Picraux in "Site Characterisation and Aggregation of Implanted Atoms in Materials". (Eds. A Perez and R Coussement Plenum Press, New York) pp307 and 325 (1980).

52. J A Borders. Ann Rev Mat Sci 9, 313 (1979).

53. J M Poate and A G Cullis in "Treatise on Materials Science and Technology", Vol 18, Ion Implantation (Ed. J K Hirvonen, Academic Press, New York) p85 (1980).

54. D K Sood. Rad Effects. (To be published 1982).

55. C M Preece, E N Kaufman, A Staudinger and L Buene in "Ion Implantation Metallurgy" (Eds C M Preece and J K Hirvonen, AIME, Warrendale, Pa) p77 (1980).

56. A D Marwick. Nucl Instrum & Meth 182/183, 827 (1981).

57. J W Mayer, B Y Tsaur, S S Lau and L S Hung. Nucl Instrum & Meth 182/183, 1 (1981).

58. B Y Tsaur, S S Lau, L S Hung and J W Mayer. Nucl Instrum & Meth 182/183, 67 (1981).

59. G Fischer, A E Hill and J S Colligon. Vacuum 28, 277 (1978).

60. I N Evdokimov and G Fischer. J Mat Sci 15 854 (1980).

61. A Gras-Marti, J J Jiminez-Rodriguez, N P Tognetti, G Carter, M J Nobes. Vacuum. (To be published 1982).

A IGNATIEV

Selective
solar absorbing coatings
by ion implantation

SYNOPSIS

Solar selective optical properties have been realized for chromium and zirconium surfaces subjected to oxygen and nitrogen ion implantation at 10, 30 and 50 keV. A surface carbon layer has been shown to be principally responsible for the optical response which shows solar absorptance of 92%.

THE AUTHOR

Dr. Ignatiev is in the Department of Physics at the University of Houston and was a visitor at the Physics Institute at Aarhus University in Aarhus, DENMARK.

INTRODUCTION

Recent interest in the utilization of solar energy has prompted development of efficient solar energy absorbing coatings. Such coatings have been vacuum evaporated or sputtered[1-3], electroplated[4,5] and chemical vapor deposited[6]. Of prime concern in developing the coatings are a high solar absorptance, $\alpha \geqslant 0.90$, low infrared emittance, $\varepsilon \leqslant 0.3$ and stability under high temperatures ($T \sim 300\text{-}600^{\circ}C$) and high solar fluxes (~ 1000 suns).

To this end and with special interest in high temperature stability, we have undertaken this study of ion implantation generation of solar absorbing coatings. Both the oxide and nitride of chromium and zirconium are quite stable at high temperatures in air and are semiconductors, thus having nominal values of solar absorptance. We have therefore chosen to implant oxygen and nitrogen ions into chromium and zirconium samples in attempts to generate efficient solar energy absorbing coatings.

EXPERIMENTAL

Sample subs of chromium and zirconium were prepared in nominal sizes of 15 x 15 x 0.5mm and were polished flat and smooth. These samples were then bombarded at the Aarhus University Ion Separator with atomic nitrogen and oxygen ions at energies of 10, 30 and 50 keV and doses of 28,000 μC, 56,000 μC and 112,000 μC. The sample chamber was held at a pressure of

5×10^{-6} torr during the implantation which at ion current densities of fractions of microamps per cm^2 required several hours of exposure per sample.

The samples were subsequently analyzed at the University of Houston to determine changes in total hemispherical spectral reflectance and atomic composition as a result of implantation. The reflectance measurements were made on a Beckman DK-2A spectraphotometer and the composition measurements were accomplished by Auger electron spectroscopy (AES)-depth profiling of the surface regions of the samples.

RESULTS

Figure 1 shows typical optical results for oxygen and nitrogen implantations in chromium and zirconium. Notice the sharp decrease in the hemispherical reflectance below 1 μm after implantation. This decrease not only increases solar absorptance over that of the base metal (to up to 92%) but its selective nature retains the high reflectance (low emittance) of the sample in the infrared. Figure 2 shows the ion energy dependence of oxygen implantation into chromium. There is lack of significant difference in the curves as a function of ion energy although there is a factor of about 1.5 to 1.6 difference in projected range of the ions[7].

Figure 3 shows the dose dependence of the hemispherical reflectance for oxygen implanted chromium. Very large changes are observed as a function of dose with increasing dosage resulting in shift of the absorber-reflector transition region to the infrared.

DISCUSSION

The relative insensitivity of the optical response to ion energy is quite surprising. In addition, the changes with dose are more severe than expected. The reflectance curves of Figure 3 have interference minima which are dependent on dose and which, if one assumes a simple 2 layer interference stack model for the surface region of the sample (Cr_2O_3 on Cr), indicate film thickness of 700Å, 1,000Å and 2,800Å for lowest to highest dose (all $\pm 15\%$ due to error in assuming an index of refraction for the overlayer). This is well above the ≈ 400Å

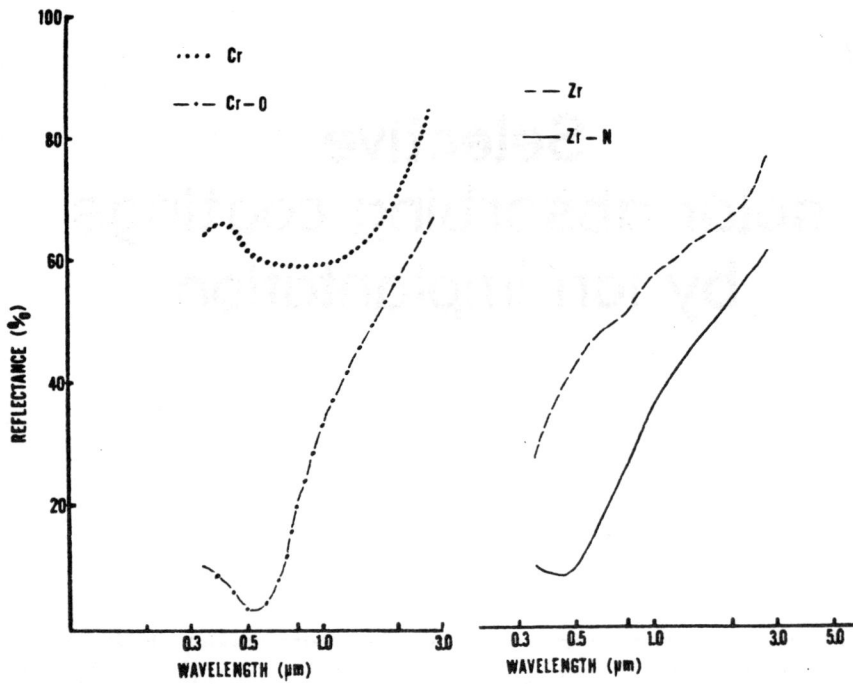

Figure 1. Hemispherical reflectance for chromium and oxygen implanted chromium (50 keV, 56,000 µC), and for zirconium and nitrogen implanted zirconium (30 keV, 28,000 µC).

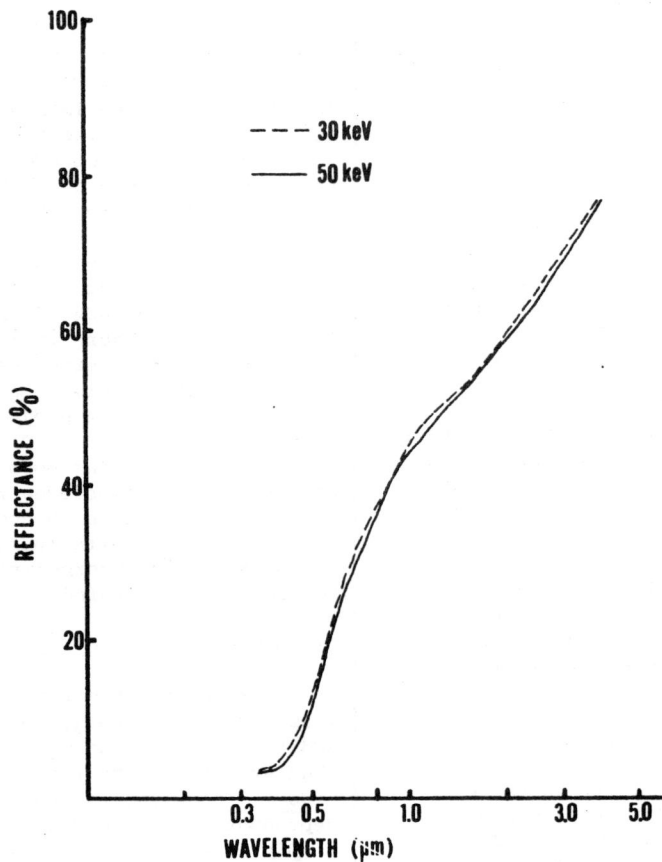

Figure 2. Hemispherical reflectance as a function of ion energy for oxygen implantation in chromium (both doses at 28,000 µC).

projected range for 30 keV oxygen ions in chromium[7].

Auger electron spectra in conjunction with inert gas depth profiling were undertaken in order to clarify these points. Figure 4 gives the depth profile of the low dose sample of Figure 3. The first striking point is that there is an immense amount of carbon at the surface of the sample. The carbon signal remains large (~90 at % and is due to graphitic carbon[8]) to an erosion time of ~9 min. From our previous experience in depth profiling metal oxide and carbide films[9] we can define an ion erosion rate of ~90 to 100Å per min. for the graphite part of the implanted sample. The carbon layer thickness is therefore ~800 to 900Å. The oxygen signal variation as a function of erosion time (depth) shows a surface enhancement and a peak at ~12-13 min. The peak is ~3 to 4 min. above the steep rise of the chromium signal at 9 min. and corresponds to oxygen deposition at a depth ~300 to 400Å below the chromium surface. This is quite consistent with the expected projected range for 30 keV oxygen in chromium.

Figure 5 gives the AES depth profile for the 50 keV oxygen implanted sample of Figure 2. As in Figure 4, a large carbon concentration is observed at the surface (equivalent in thickness to that in Fig. 4) and an oxygen peak is located at the surface of the sample as well as a second peak at 7 to 8 min. after the chromium surface. This second peak corresponds to oxygen atoms deposited 700 to 800Å below the surface of the chromium and is consistent with the expected 700Å projected range of 50 keV oxygen ions in chromium[7].

Figure 6 shows the AES depth profiling results for a chromium sample implanted with oxygen at 50 keV and 56,000 μC. The optical response of this sample is essentially identical to the 56,000 μC sample of Figure 3. Figure 6 again shows the 2 peak structure for oxygen with the second peak at about 700 to 800Å below the chromium surface. The surface localized carbon, however, is now <u>thicker</u> (by almost 50%) than in the 28,000 μC sample of Figure 5. It is now clear that the changing optical response of Figure 3 is simply due to the growth of a thicker carbon (graphitic) layer on the sample. This is most probably due to the fact that the sample had to be bombarded for a longer time to obtain the larger ion dose, and this was done under relatively poor vacuum conditions. A residual gas analysis was not available for the vacuum chamber; however, the gases are expected to be principally hydrocarbons (from diffusion pump and fore pump oils) and water vapor. The ion beam therefore probably cracked the surface adsorbed hydrocarbons and as a result plated a thick graphite layer on the samples. This graphite layer, which does not form in absence of the ion beam, is, however, very tenacious and although no mechanical adhesion tests were done, the samples were found to be very resistant to flaking, peeling and scratching. This mechanical stability is a very positive factor for a viable solar absorbing coating.

Further work is currently underway to understand the exact nature of this carbonaceous layer and to more clearly delineate the role of the implanted atoms (versus the carbon layer) in defining the optical response of the system. To this end, short-time, large-dose exposures of chromium samples to oxygen ions are being planned.

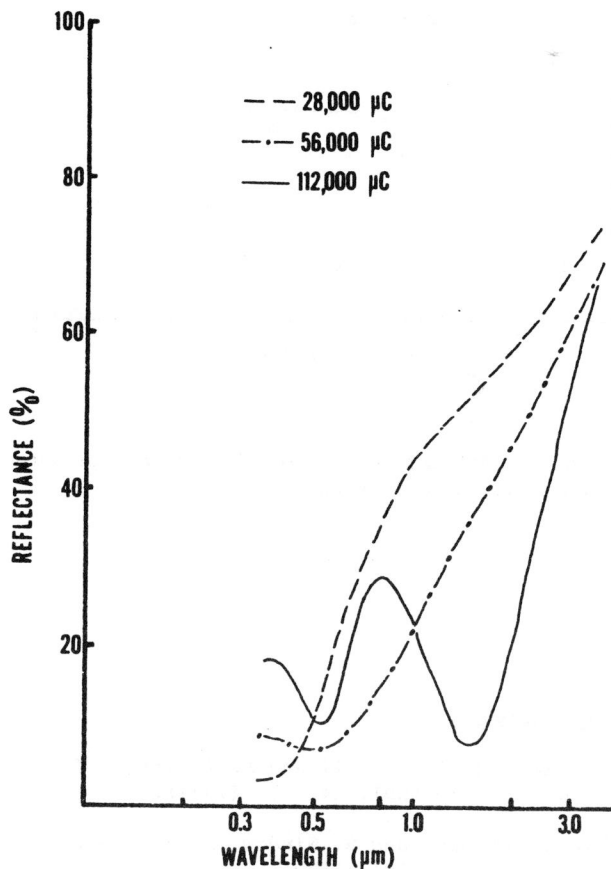

Figure 3. Hemispherical reflectance as a function of ion dose for 30 keV oxygen implantation in chromium.

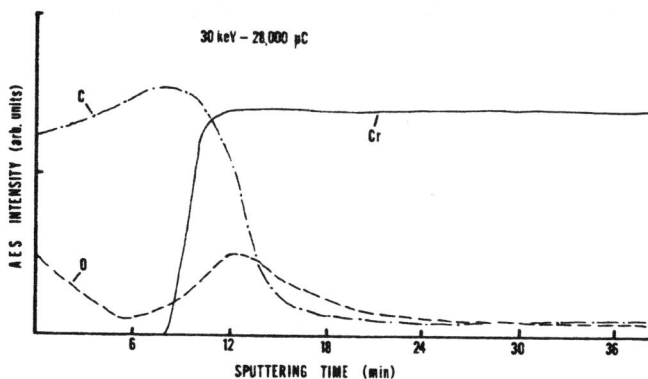

Figure 4. AES depth profile of 30 keV oxygen implanted chromium.

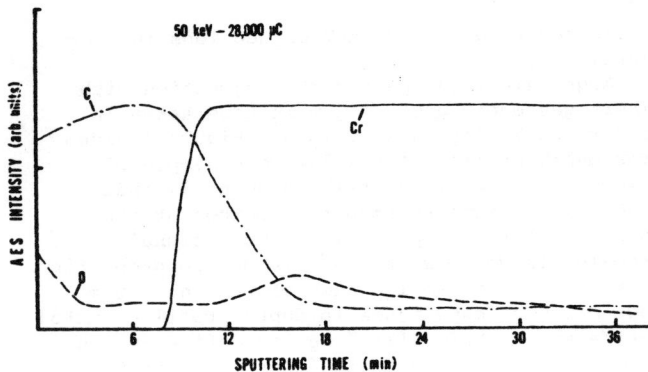

Figure 5. AES depth profile of 50 keV oxygen implanted chromium.

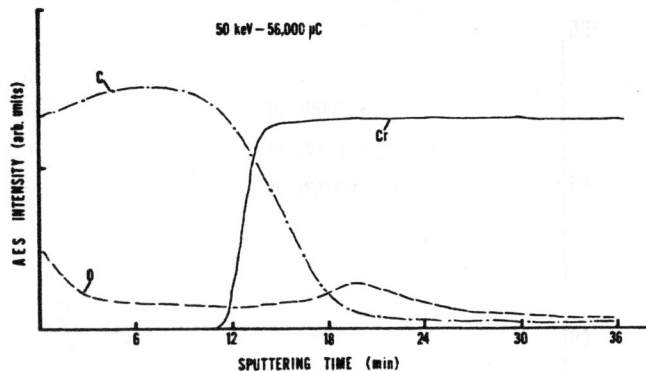

Figure 6. AES depth profile of 50 keV high dose oxygen implanted chromium.

CONCLUSION

Samples of chromium and zirconium have been implanted with oxygen and nitrogen ions of 10, 30 and 50 keV energy and doses of 28,000, 56,000, and 112,000 µC. The optical properties of samples were changed significantly only as a result of ion dose. This was determined to be due not to increased oxygen in the substrate but to an increase in the thickness of a tenacious graphite layer which was plated on the surface of the sample from the interaction of the ion beam with the hydrocarbons adsorbed on the surface of the sample from the vacuum chamber residual gas. Solar absorptance increases of over 100% have been obtained for treated chromium samples with final absorptances of 92% being achieved.

ACKNOWLEDGMENTS

The assistance of the staff at the Physics Institute, Aarhus University, with special assistance from G. Sørensen, is gratefully acknowledged. This work has been supported in part by the U. S. Department of Energy.

REFERENCES

1. C. Doland, P. O'Neill and A. Ignatiev, J. Vac. Sci. Technol. 14, 259 (1977).

2. G. L. Harding and B. Window, J. Vac. Sci. Technol. 16, 2101 (1979).

3. H. G. Craighead, R. Bartynski, R. A. Buhrman, L. Wojcik and A. J. Seivers, Sol. Energy Mater. 1, 105 (1979).

4. A. Ignatiev, P. O'Neill and G. Zajac, Sol. Energy Mater. 1, 69 (1979).

5. G. B. Smith, A. Ignatiev and G. Zajac, J. Appl. Phys. 51, 4186 (1980).

6. K. A. Gesheva, K. Seshan and B. O. Seraphin, J. Appl. Phys. 52, 1356 (1981).

7. K. B. Winterbon, Ion Implantation Range and Energy Deposition Distributions, (IFI/Plenum, New York, 1975).

8. T. Haas, J. Grant and G. Dooley, J. Appl. Phys. 43, 1855 (1972).

9. G. Zajac, G. B. Smith and A. Ignatiev, J. Appl. Phys. 51, 5544 (1980).

M IWAKI, Y OKABE, H HAYASHI, A KOHNO, K TAKAHASHI, and K YOSHIDA

Surface characteristics of ion-implanted steels

SYNOPSIS

Studies have been made of the surface-layer characteristics of ion-implanted steels. Targets are commercial mild steels. Ion implantations have been performed with doses of 10^{15}-10^{17}/cm^2 at energies of 50-150 keV. Implanted elements are mainly nitrogen and chromium, and additionally oxygen, boron, silicon, argon, titanium, nickel and copper. The surface-layer characteristics of implanted steels have been investigated from the point of view of friction, hardness, wear and corrosion. These results suggest that ion implantation is a useful technique for the control of surface characteristics and especially for the improvement of wear and corrosion resistance.

THE AUTHORS

are in The Institute of Physical and Chemical Research, Hirosawa, Wako-shi, Saitama, Japan.

INTRODUCTION

Ion implantation has been successfully used in the fabrication of semiconductor devices as a method of doping controlled amounts of impurities. For some ten years the technique has been applied to various metallic targets in order to study their mechanical and chemical properties such as friction, wear, corrosion, etc.[1] Many results have shown that ion implantation in metals with a high dose is a useful technique for the formation of a new metastable surface alloy structure of controlled composition without many of the limitations of thermal alloying.

In the case of ion implantation in steels, nitrogen of the gaseous elements and chromium of the metal elements are considered as important implanted ions for the improvement of wear and corrosion resistance. Nitrogen implantation would be carried out for the purpose of combining it with iron, e.g. forming an iron nitride. Moreover, high intensity of nitrogen beam current can be easily obtained by a simple implanter,[2,3] and therefore nitrogen implantation in steels may be put to practical use. In the case of chromium implantation, the corrosion behavior of implanted steels was usually compared with that of conventional stainless steels.[4,5] Implanted chromium may be easily combined with oxygen to form a stable oxide layer near the surface and consequently the corrosion resistance of steel may be improved.

The authors have investigated ion implantation of metals for the purpose of applying to their surface a layer of alloy. Ion implantation with a high dose greatly changes the surface characteristics of metals. Ions were implanted into steel targets and the friction characteristics, surface hardness, wear resistance and electrochemical properties of implanted steels were investigated. In this report these properties are described, with special attention paid to nitrogen and chromium implantation in steels.

EXPERIMENTAL PROCEDURE

2.1 Ion Implantation

Substrates used were pure iron sheets or commercially available mild steel plates and carbon steels. Before ion implantation, the specimens were usually mirror-polished by a buffing wheel and ultrasonically cleaned in trichloroethylene. Part of the sample was masked with aluminum sheet for measuring the relative change in surface characteristics, e.g. for a clear distinction between the friction coefficients of implanted and unimplanted regions. Implanted elements were nitrogen, oxygen, boron, silicon, argon, titanium, chromium, nickel and copper. Ion implantations were performed with doses of 10^{15}-10^{17} ions/cm^2 at energies of 50-150 keV at a pressure of about 2×10^{-4} Pa by using the RIKEN 200 kV Low Current Implanter.[3] The target temperature during ion implantation rose from room temperature up to about 200°C.

(a) P = 100 gf

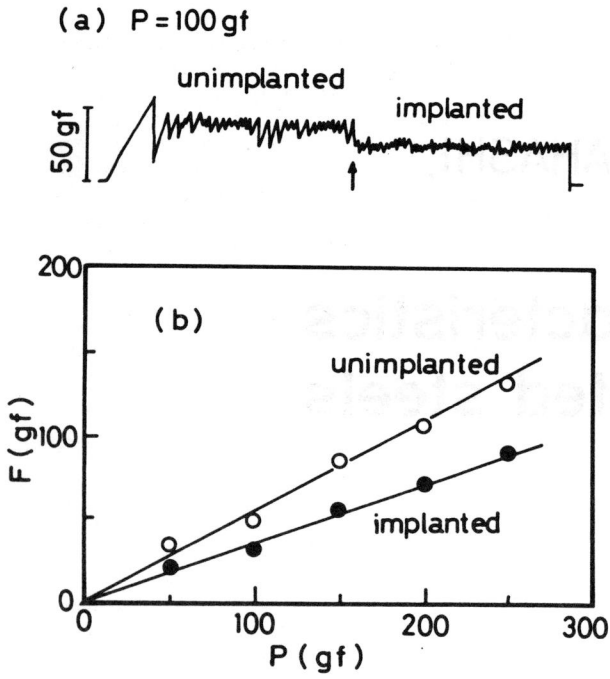

Fig. 1. Friction trace (a) and relationship (b) between friction force (F) and normal load (P) for steel implanted with $5 \times 10^{16} N_2/cm^2$ at 100keV.

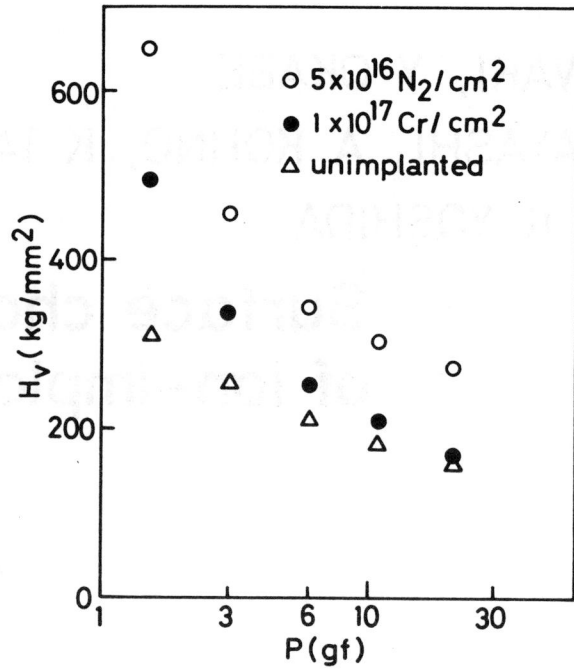

Fig. 3. Vickers hardness (H_V) as a function of normal load (P) on unimplanted, nitrogen- and chromium-implanted steels.

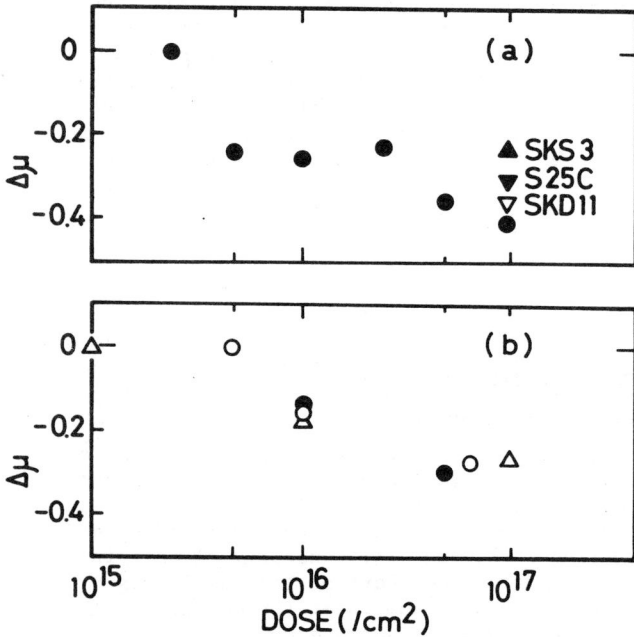

Fig. 2. Relative change in friction as a function of dose on nitrogen (a) and chromium (b) implanted steels at 50(○), 100(●) and 150(△,▲,▼,▽)keV.

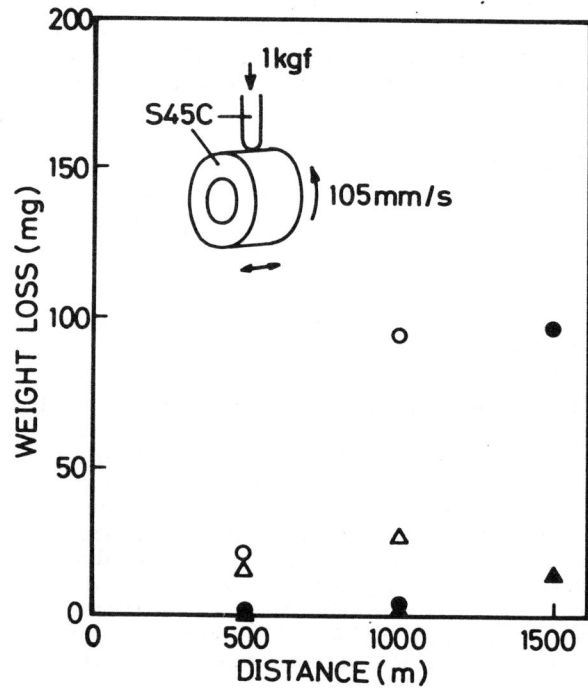

Fig. 4. Relationship between amount of wear and sliding distance in the pin(△)-unimplanted cylinder(○) and the pin(▲)-implanted cylinder (●).

2.2 Friction Tests

The friction coefficients were measured by using the Bowden-Leben type friction testing machine.[6] One of two contacting materials was always the unimplanted rigid bar, whose top was a carbon steel (JIS, S45C) with a half sphere of diameter 4 mm. The normal load applied to the pin ranged from 50 to 250 gf. The pin was slid on the surface continuously from unimplanted to implanted regions at a velocity of 2.7 mm/s. The friction force was measured by a strain gauge fastened on the spring plate. All the friction tests were carried out under unlubricated conditions at atmospheric room temperature.

2.3 Hardness Tests

The Vickers hardness of specimens was measured by using the micro-Vickers hardness tester. The normal load on both unimplanted and implanted regions of specimens was in the range 1.5-100 gf for 15 or 30 seconds. Materials used in this experiment were the same as specimens in the friction measurements.

2.4 Wear Tests

Wear tests were carried out by using the pin-cylinder configuration test equipment. The effect of ion implantation on wear resistance was estimated by weight loss of implanted and unimplanted cylinders before and after wear tests. One of two contacting materials was always the unimplanted pin, which was made of a carbon steel (S45C) with a radius of 2.5 mm. The test-pin was kept in contact with the side of a cylinder, which was rotated continuously at a constant velocity. The friction force was measured by the load cell on the arm supporting the pin. The amount of wear was measured by the loss in weight before and after wear tests.

2.5 Corrosion Tests

The aqueous corrosion behavior of these specimens was electrochemically investigated by means of a conventional voltammetric system with a three electrode type cell, which was employed in multi-sweep cyclic voltammetry with a 0.5 mol/dm^3 acetate buffer solution of pH 5.0. All the measurements were made at 25.0+0.1°C. The cell construction and the instrument for the experiment were given in detail in ref. 7.

RESULTS

3.1 Friction

All the high dose ion implantations made a change in friction. A typical friction trace of a nitrogen molecular implanted steel plate is illustrated in Fig. 1 (a). Ion implantation was performed with a dose of 5×10^{16} N_2/cm^2 at 100 keV. The arrow indicates the boundary between implanted and unimplanted regions. The magnitude of the friction force is obviously different between unimplanted and implanted regions, and the stick-slip phenomenon becomes weak after the nitrogen implantation. The groove of the friction track was 100 times deeper in the unimplanted layer.

The friction forces were measured as a function of the normal load, as shown in Fig. 1 (b). In both unimplanted and implanted regions, the friction force is directly proportional to the applied load in the range of 50-250 gf. The friction coefficient, defined as the ratio of the friction force to the normal load, is 0.54 on the unimplanted region and 0.35 on the nitrogen implanted region of the specimen indicated in Fig. 1 (a).

All of the friction coefficients of the unimplanted region were 0.5-0.6, and differed slightly from specimen to specimen. To estimate the effect of ion implantation on friction, the relative change in friction is defined as

$$\Delta\mu = (\mu_i - \mu_u)/ \mu_u$$

where μ_i and μ_u are the friction coefficients on implanted and unimplanted regions, respectively.

The relative change in friction is shown as a function of the total dose at energies of 50, 100 and 150 keV in Fig. 2, where (a) and (b) indicate nitrogen molecule and chromium implanted specimens respectively. As the dose of nitrogen molecular ions increases above 5×10^{15}/cm^2, the friction coefficient decreases markedly. The friction coefficient of tool steels implanted with a dose of 1×10^{17} N_2/cm^2 also shows a tendency to decrease and their relative change in friction is -0.2 to -0.4, as shown in the figure.

In the case of chromium implantation, an increase in the total dose resulted in a decrease of the friction coefficient, which was independent of the acceleration energy.

3.2 Hardness

The Vickers hardness is shown as a function of the applied normal load in Fig. 3. Ion implantations were carried out at an energy of 150 keV with doses of 5×10^{16} N_2/cm^2 and 10^{17} Cr/cm^2. Even in the case of the unimplanted specimens, decreasing the normal load applied for measurements results in an apparent increase in hardness. A hardening phenomenon at light loads may be caused by work hardened layers at the surface, be because the information of characteristics near the surface can be obtained at the lighter load.

Nitrogen implantation makes the surface harder by a factor of two at the load of 1.5 gf. Even if a load of 21 gf is applied on the implanted layer, it appears harder than the original material. The hardness of the nitrogen implanted specimen is at least 1.4 times that of an original specimen at the normal load of 21 gf. The hardening effect of the nitrogen implantation was also observed in the case of tool steels (JIS S25C, S45C, SKS3 and SKD11). All the relative changes in hardness were 0.4 in the case of the implantation with 10^{17} N_2/cm^2 at 150 keV.

With chromium implantation, the surface layer becomes hard at the low load, but an applied load above 11 gf removes the difference between the hardness of implanted and unimplanted specimens. At the normal load of 1.5 gf, the chromium implanted specimen is 1.5 times as hard as the unimplanted specimen.

3.3 Wear

The substrate used as an implanted target probe was a cylinder of 40 mm in diameter and 25 mm in length. Ion implantation for wear tests was performed with the ion beam produced from

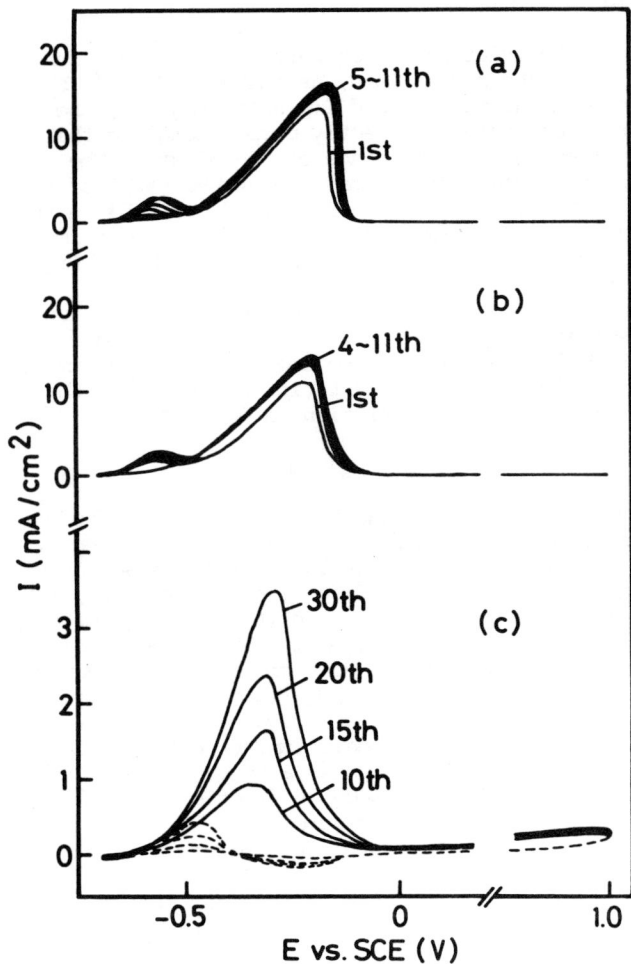

Fig. 5. Typical multi-sweep cyclic voltammograms for unimplanted(a), nitrogen implanted(b) and chromium implanted(c) steels.

(a) unimplanted

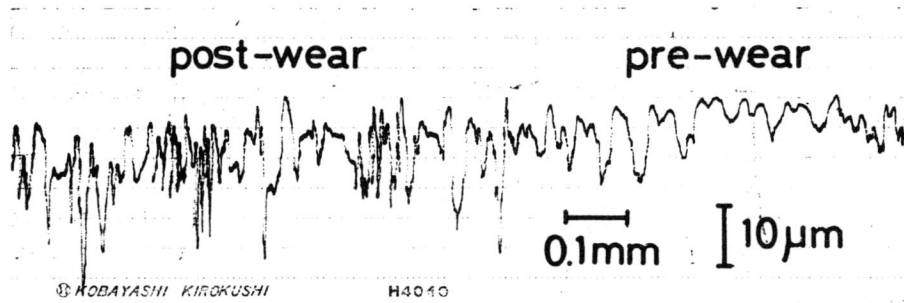

(b) implanted

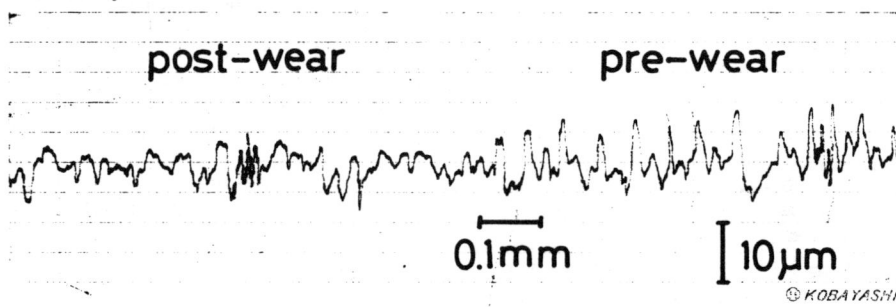

Fig. 6. Surface roughness for unimplanted (a) and nitrogen implanted (b) steels before and after wear testing at the sliding distance of 500 m.

nitrogen gas. The flux of the ion beam was 0.0160 C/cm², The ion beam contained 32%N, 48%N_2 and 20%NO, which was measured by mass analysis before ion implantation. In this case the content of nitrogen atoms in the beam is about 90%.

Figure 4 shows the relationship between the amount of wear loss and sliding distance, when the pin was oscillated along the side of a rotating cylinder. The applied normal load was 1 kgf and the sliding velocity was 105 mm/s. As shown in the figure, the amount of wear is reduced by ion implantation. The difference in weight loss is remarkable below a sliding distance of 1000 m. Typically at a sliding distance of 500 m, loss in weight of the nitrogen implanted cylinder is 100 times smaller than that of the unimplanted cylinder.

In the case of the pin, the amount of wear is different whether contacting materials are unimplanted or implanted cylinders: loss in weight of the pin in contact with the implanted cylinder reduces by a factor of 20, as compared with that in contact with the unimplanted cylinder. The result shows that the nitrogen implanted material affects the wear property of the other unimplanted specimen in contact with it. Nitrogen implantation in steels results in an improvement in wear resistance even under unlubricated conditions.

3.4 Corrosion

Nitrogen and chromium implanted steels showed an improvement in corrosion resistance in the atmosphere at room temperature. The aqueous corrosion investigated by means of electrochemical measurements was different from the corrosion phenomenon at atmospheric room temperature.

Figure 5 shows the multi-sweep voltammograms obtained for unimplanted (a), nitrogen molecule implanted (b) and chromium implanted steels (c). In the figure, the anodic current peaks were observed in the potential region of -0.7 to -0.2 V vs. a saturated calomel electrode (SCE), which corresponds to the anodic dissolution behavior of the metal electrodes. The potential region of 0 to 1.0 V vs. SCE, in which the current depressed, corresponds to the passivation of the electrode surfaces. The anodic peak current increases with increasing number of potential sweep cycles, and such multi-sweep voltammograms show that the surface layer is gradually dissolved with repetition of the potential sweep.

The behavior of voltammograms of the nitrogen implanted steel is almost the same as that of the unimplanted specimen. Nitrogen implantation therefore, did not improve the aqueous corrosion resistance of steels in this experimental condition.

With chromium implantation in steels, the anodic peak current density decreases compared with that of the unimplanted specimen. The electrochemical properties of the chromium implanted steel are almost the same as those of Fe-18% Cr alloy (JIS SUS 430), but the repetition of potential sweep results in increasing the anodic peak current which may be caused by the local dissolution of the material surface layers.

DISCUSSION

The friction coefficient was dependent on the implanted elements. As indicated in Fig. 2, nitrogen and chromium implantations result in decreasing the friction coefficient of steels. The same phenomenon was also obtained by boron and oxygen implantations in steels. In these cases, the friction coefficient becomes low as the total dose increases.

Copper or nickel ion implantation in steels caused an increase in friction coefficient. In this case, the friction coefficient showed a tendency to increase as the total dose at each energy became larger. It also increased as the acceleration energy at the same dose became lower. The depth contributed to the friction coefficients was investigated in the relationship with the depth profile of implanted elements to be the thin layer from the surface to 40 nm.[6]

The relationship between friction and hardness properties was presented by E.P. Bowden and D. Tabor.[8] The friction coefficient is written as the ratio of two quantities τ and p, respectively representing the resistance to plastic flow of the weaker of the contacting materials in shear and in compression, as follows:

$$\mu = F/P = A_r\tau/A_r p = \tau/p$$

In this equation, F and P are the friction force and the applied normal load, respectively; A_r is the real area of the contact; p is approximately equal to the hardness. Considering the equation, the friction coefficient decreases as the hardness increases.

The Vickers hardness of steels increased in the case of nitrogen or chromium implantation with a high dose, when the friction coefficient measured with the same samples became low. The result suggests that the decrease of the friction coefficient may be caused by the hardening effect of nitrogen or chromium implantation.

No change of the hardness of argon, nickel and copper implanted steels was found at any normal load. If the surface layer becomes soft by ion implantation, the measurement of the surface hardness is strongly affected by a bulk hardness and therefore there seems no change in hardness. In view of the increase of friction coefficients, the implanted layers may be softened by argon, nickel and copper implantations.

Wear resistance of steels under unlubricated conditions at room temperature was improved by nitrogen implantation, which increased the hardness as the friction coefficient decreased. The hardness of the nitrogen implanted cylinder used for wear tests was 1.4 times as large as that of the unimplanted cylinder. The friction coefficient during wear testing was 0.35 in the system of the pin-implanted cylinder and 0.55 in the pin-unimplanted cylinder. Its relative change was -0.36, which is almost the same as the value indicated in Fig. 2.

Figure 6 shows typical roughness of cylinders which was measured at the sliding distance of 500m in Fig. 4. The surface roughness height before wear testing is 5 μm at centre line average. After wear testing, the surface is as rough as 15 μm in the case of the unimplanted cylinder and under 5 μm in the case of the implanted cylinder. This depth is about 50 times as deep as the ion range.

Even if the pin was slid on the same area of a cylinder an improvement of wear resistance was observed.[3] In this experiment, contacted surfaces are considered to be stable up to 100 m, where loss in weight of the unimplanted cylinder was 10 times larger than that of the implanted cylinder. Nitrogen implantation in steels results in greatly improving the wear resistance without lubrication. In this experiment, it is also found that the implanted material affects the wear property of the other unimplanted specimen in contact with it.

Nirogen implantation caused the corrosion resistance to improve at atmospheric room temperature, but did not change the aqueous corrosion resistance. Nickel and copper implantations with a high dose slightly improved the aqueous corrosion resistance of steels. In the case of chromium implantation with a low dose, the surface layer is dissolved by the oxidation reaction during the multi-potential sweeps.

Chromium implantation with a high dose results in improving the aqueous corrosion resistance and exhibits almost the same property as the conventional alloy of Fe-18% Cr. In order to see the effect of electrochemical reactions of the electrode surface on the composition of a surface layer by voltammetric measurements, the concentration profiles were measured by secondary ion mass spectroscopy. The results show that the multi-sweep voltammetric measurement did not affect the concentration profile of chromium implanted in steels with a dose of 10^{17} Cr/cm^2. However, it seems that the increase of the anodic peak current with repetition of sweep cycles, as shown in Fig. 5, is caused by local corrosion which was observed by SEM.

Silicon implanted steels have the same properties as chromium implanted steels. In the case of titanium ion implantation, its electrochemical property was more inert than that of chromium implanted steels.

CONCLUSION

Ion implantation in steels was carried out with high doses to investigate the friction, hardness, wear and corrosion behavior. Main implanted elements were nitrogen and chromium and auxiliary elements were boron, oxygen, silicon, argon, nickel and copper. These results were as follows.

(1) The friction coefficients of steels decrease in the case of chromium, oxygen, nitrogen and boron implantations and increase in the case of copper and nickel implantations.

(2) Nitrogen and chromium implanted steels become harder than unimplanted steels.

(3) Nitrogen implantations improve the wear resistance under unlubricated conditions at atmospheric room temperature.

(4) Chromium implanted steels have a high resistance to corrosion. The elements which improve corrosion resistance are chromium, silicon, titanium, nickel and copper.

REFERENCES

1 J.K. Hirvonen: Ion Implantation, Treatise on Materials Science and Technology, Vol. 18 (Academic Press, New York, 1980).

2 G. Dearnaley: Materials in Engineering Applications 1 (1978) 28-41.

3 M. Iwaki, Y. Okabe, S. Namba and K. Yoshida: Nucl. Instrum. & Methods 189 (1981) 155-159.

4 V. Ashworth, D. Baxter, W.A. Grant, R.P.M. Proctor and T.C. Wellington: Ion Implantation in Semiconductors and Other Materials, ed. S. Namba (Plenum Press, New York, 1975) 367-375

5 Y. Okabe, M. Iwaki, K. Takahashi, S. Namba and K. Yoshida: Surf. Sci. 86 (1979) 257-263.

6 M. Iwaki, H. Hayashi, A. Kohno and K. Yoshida: Jpn. J. Appl. Phys. 20 (1981) 31-35.

7 K. Takahashi, Y. Okabe and M. Iwaki: Nucl. Instrum & Methods 182/183 (1981) 1009-1015.

8 E.P. Bowden and D. Tabor: The Friction and Lubrication of Solids (Oxford University Press, Pt. II, 1964).

E RAMOUS, G PRINCIPI, L GIORDANO, S LO RUSSO, and C TOSELLO

Thermal effect of nitrogen implantation on high carbon steels: influence of current density

SYNOPSIS

Quenched and annealed samples of a high carbon steel implanted with 105 keV nitrogen at the 3×10^{17} ions/cm^2 nominal dose and current densities of 50,100 and 200 μA/cm^2 have been analysed by a nuclear technique to measure the retained dose which resulted, dependent on the beam current density as well as on the steel starting structure. Microhardness measurements and Conversion Electrons Mössbauer Scattering (CEMS) have been used to study the structure modifications induced by the different implantation conditions. The mean implantation temperature of the samples, as evaluated on the basis of the microhardness values and checked by a thermocouple, lies in the tempering range of hardened structures at the above implantation current densities.

THE AUTHORS

E. Ramous, G. Principi and L. Giordano are with the Istituto di Chimica Industriale, Università di Padova (Italy), S. Lo Russo is with the Istituto di Fisica, Università di Padova (Italy) and C. Tosello is with the Istituto per la Ricerca Scientifica e Tecnologica, Povo, Trento (Italy)

INTRODUCTION

It has been reported recently that nitrogen implantation on steels improves some technological properties like wear, fatigue and corrosion resistance[1-6]. Radiation damage, interstitial-rich region formation and local precipitation of nitrides have been invoked to explain these effects.[2]

The surface compounds formation (nitrides and carbonitrides) at nominal doses in the range $1 \div 8 \times 10^{17}$ N$^+$/cm^2 has been observed by means of Conversion Electrons Mössbauer Scattering (CEMS) for high and medium-carbon steels[7,8] and for chromium steels[9]. Matrix composition and microstructure play a primary role in the sequence of formation of these compounds, by increasing the dose. It has been shown, for example, that in martensitic steels ε-iron phases are formed, whereas in tempered martensite the γ'-Fe$_8$N nitride also appears[8], as well as in the case of pure iron[10].

In many cases discrepancies have been noted between the nominal dose and the dose measured by nuclear techniques. Several reasons may be advanced: the molecular beam component due to implantation carried out without mass analysis, sputtering effects, composition and structure of the matrix, sample heating due to the energy transferred by the beam, and so on.

The knowledge of the temperature during irradiation is important when the sample may undergo thermal effects like over-tempering. These effects may be limited when implanting at high dose-rate for example by moving continuously large samples or a batch of samples through the ion beam. On the other hand, tempering of quenched steels may be used for an indirect evaluation of the temperature. The main purpose of this work is to point out the heating effect during nitrogen bombardment at high implantation dose-rate, in connection with the measured dose of implanted atoms, for steels with different starting structure.

EXPERIMENTAL

Disks of 25 mm diameter and 4 mm thickness of UNI C85 high-carbon steel (in wt%: C=0.85; Mn=0.3; Si=0.3), a material generally employed as tool steel for which hardening treatments are customary, were heated at 780 °C for 15 min and subsequently quenched in water. Other disks of the same steel were annealed at 700 °C for 3 h. The surfaces of the samples were mechanically polished before implantation, in order to remove the possible decarburized

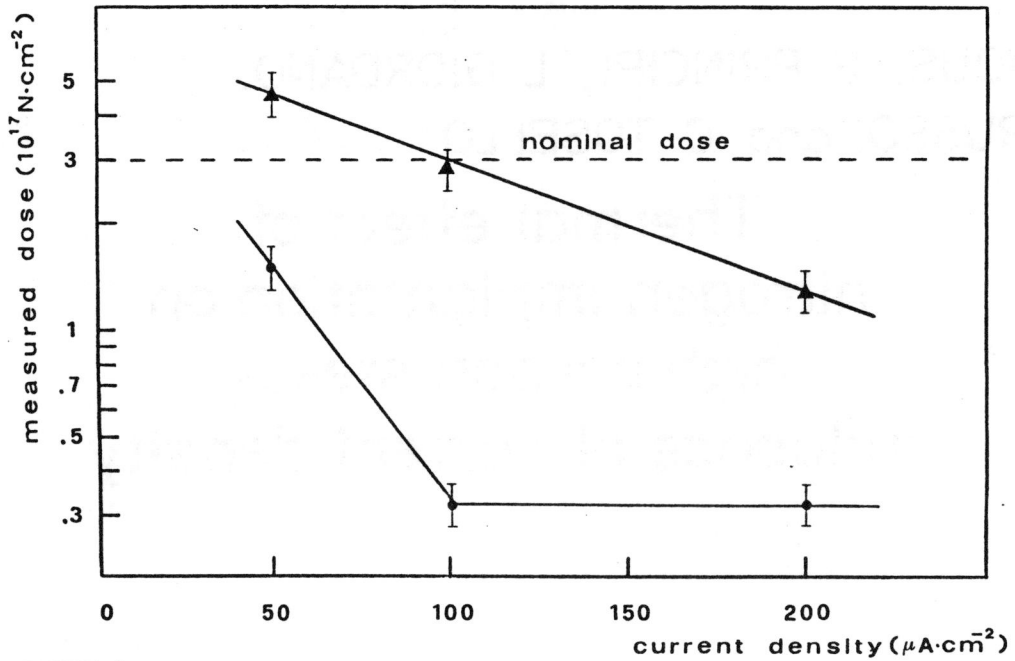

FIGURE 1

Measured dose as a function of the current
density for annealed (▲) and quenched (●)
nitrogen implanted C85 steel

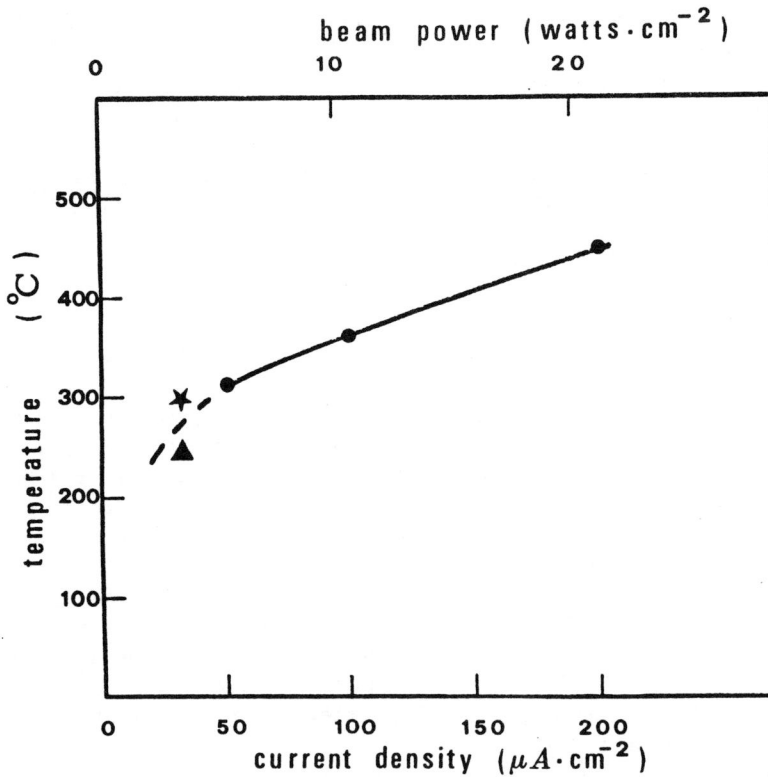

FIGURE 2

Evaluated implantation temperature as a function
of the beam current density (105 keV nitrogen
ions) and of the corresponding transferred beam
power: (●) this work, (▲) evaluated from
the microhardness profile of quenched samples
of ref. 9, (★) ref. 12 (as upper limit)

11.2

surface layer.

Nitrogen implantation has been performed by using the 120 kV accelerator of I.R.S.T. (Trento) provided with an Ortec duo-plasmatron ion source suitable for high current. Due to the absence of mass analyzer, an admixture of N^+ and N_2^+ was present in the beam, which obviously affected the implantation profile. The steel specimens were implanted at fixed energy (105 keV) and nominal dose (3×10^{17} N ions/cm^2) with three different current densities (50, 100 and 200 μA/cm^2). The disks were in thermal contact with the sample holder and some of these were connected with a thermocouple for a direct evaluation of the mean implantation temperature.

Nuclear reaction analysis was performed on the implanted samples to give a quantitative trend of the retained dose in the near surface region as a function of the implantation current density. The $^{14}N(d,p_c)^{15}N$ nuclear reaction induced by a 610 keV deuteron beam from the 2 MeV Van de Graaff accelerator at the Laboratori Nazionali di Legnaro was used for this purpose.

The microhardness tests were performed with a pyramidal indenter under a load of 200 g.

The CEMS spectra were obtained with a conventional spectrometer using a gas flowing (95% He-5% methane) proportional counter and a source of ^{57}Co (nominal activity 100 mCi) in a Rh matrix. A least-squares minimization routine was employed to compute hyperfine parameters and relative areas of the spectral components. CEMS information is relevant to a depth not more than 200 nm below the surface.[10]

RESULTS AND DISCUSSION

Retained dose measurements

We define the above mentioned "near surface region" as a depth of about 200 nm below the sample surface. This is the depth analyzed by the described nuclear technique in the chosen experimental conditions and also corresponds approximately to the depth analyzed by the CEMS technique. Obviously the measured dose in the near surface region corresponds to the total retained dose only in the absence of nitrogen migration at depths >200 nm. If nitrogen is present at greater depth, the reported values are slightly overestimated if referred to the near surface region. This fact enhances the different behaviour of quenched and annealed steel shown in Fig. 1, where the measured retained dose is reported as a function of the implantation beam current density. A very different behaviour is evident between the annealed and the quenched samples: the retained dose for the annealed samples decreases from 4.5×10^{17} N ions/cm^2 to 1.3×10^{17} N ions/cm^2, whereas it is systematically and surprisingly lower for the quenched samples and seems to be independent from the current density over about 100 μA/cm^2.

The measured dose for the annealed steel implanted at the lowest current density is 50% higher than the nominal dose. This can be attrib-

uted to the molecular component of the unanalysed nitrogen beam. The error of the reported nuclear measurements can be estimated within 10%, as shown in the figure. In order to explain the different behaviour of annealed from quenched steels, temperature evaluation during implantation and structural analysis of the implanted layers has been performed.

Microhardness tests and temperature evaluation

The microhardnesses measured at the implanted and unimplanted sides of the studied samples are reported in table 1 together with the values measured on unimplanted samples. No strong difference is observed between the two sides of annealed samples, since the implantation thermal effect cannot induce any further modification of the bulk structure. The slightly higher values for the implanted side cannot be considered a quantitative evaluation of the hardness improvement induced in the implanted layer, because the indenter penetration (some microns) is deeper than the implantation range (less than 200 nm). However, the obtained results demonstrate that nitrogen implantation produces an increase of the surface hardness great enough to be detected also by a method which tests essentially the unimplanted zone underlying the implanted one. The microhardnesses of the unimplanted sides are obviously the same, within the error, of that measured on the unimplanted sample (185 ± 5 Vickers). Possible dependence of the microhardness upon the beam current density cannot be detected.

Different is the case of quenched samples where the microhardness of the implanted and unimplanted sides vary with the beam current density. The values for the unimplanted sides, lower than for the unimplanted samples ($\cong 960$ Vickers) and decreasing with current density, are a direct evidence of the thermal effect of implantation.

We can say that ion implantation on quenched samples actually produces both doping and tempering effects, within the used range of beam current density. The temperature reached during implantation can be roughly estimated from the tempering behaviour of the considered quenched steel. The estimated values, also reported in table 1, are plotted in Fig.2, together with other data from literature, as a function of the beam current density. The mean sample temperature is dependent on geometrical factors like shape and size of the workpiece, ratio of the beam section area to the overall implanted surface, and so on. Owing to the thermal conductivity of a metallic sample and the very long implantation time (several hours), the temperature of the sample during the treatment is to be considered nearly uniform, except for the thin layer where the implanted ions interact with the matrix atoms.

The reported values are obviously relevant to our samples, of particular shape and size, and to our implantation device. Anyway, we may assert that the heating effect by implanting in

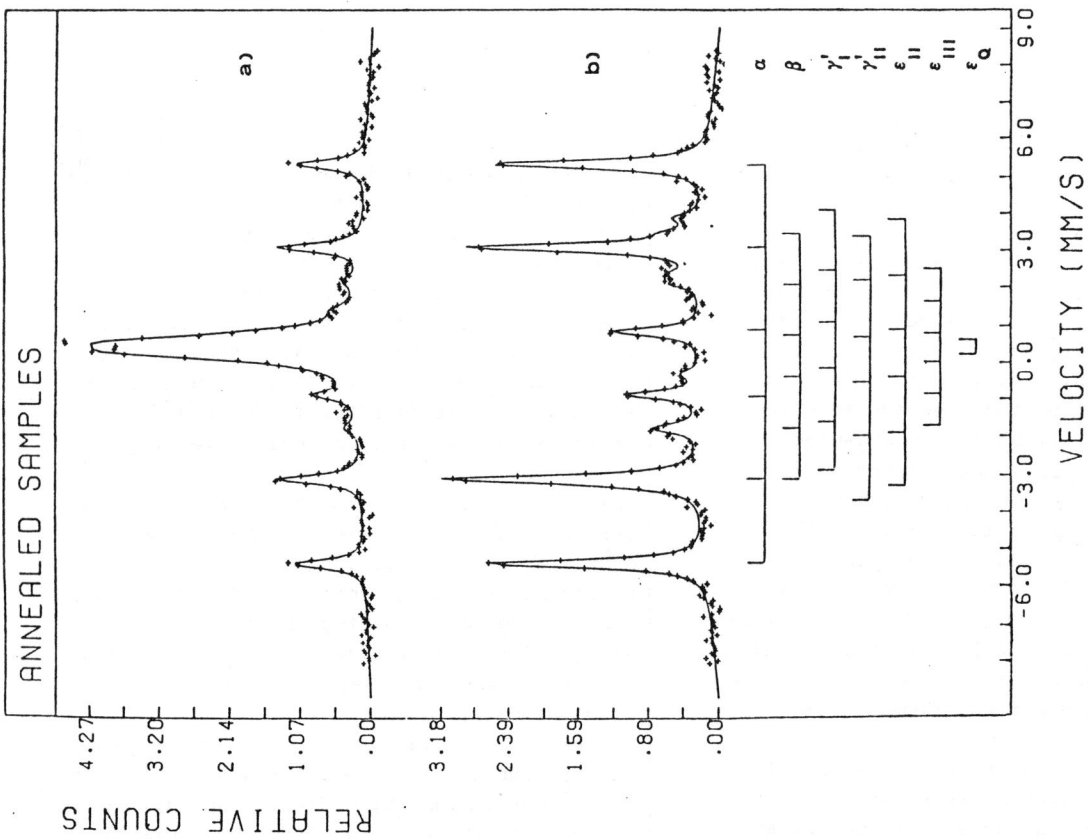

FIGURE 4

Mössbauer backscattering spectra of the quenched samples implanted at 50 μA/cm^2 (a) and at 200 μA/cm^2 (b) current densities. Components are as in Fig. 3

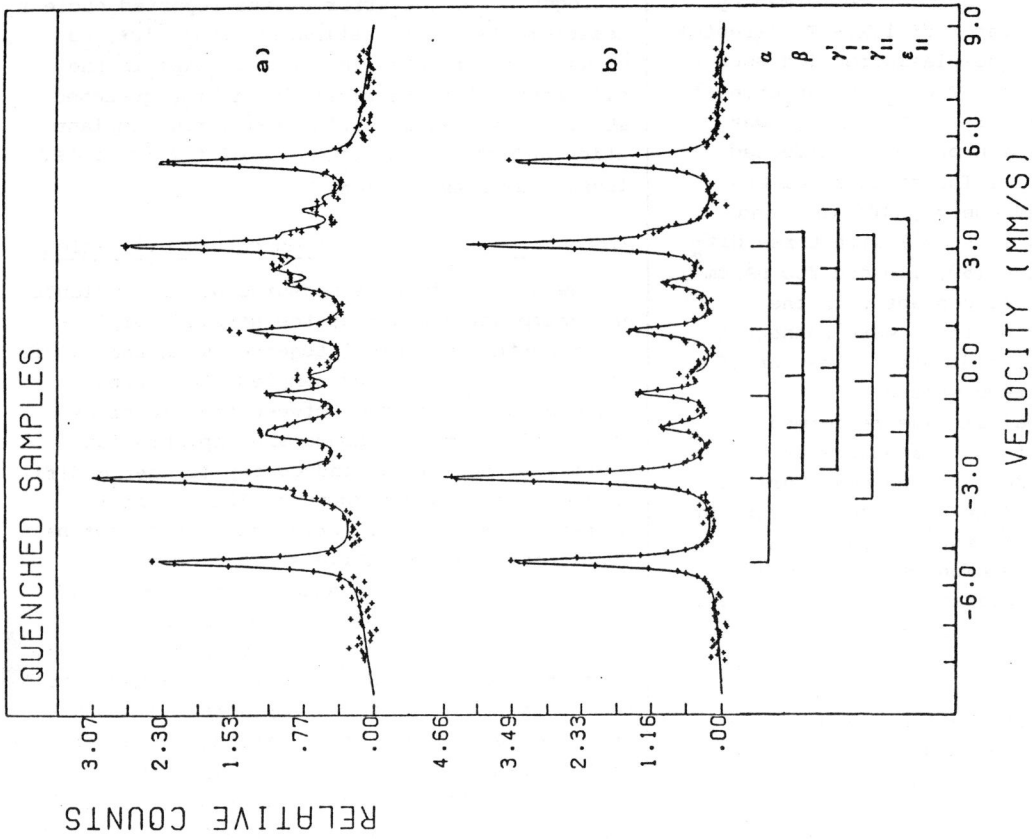

FIGURE 3

Mössbauer backscattering spectra of the annealed samples implanted at 50 μA/cm^2 (a) and at 200 μA/cm^2 (b) current densities. Components are: α ferrite, β cementite, γ'$_I$ and γ'$_{II}$ Fe$_4$N nitride, ε$_{II}$, ε$_{III}$ and ε$_Q$ Fe$_{2+x}$N carbonitride

the considered range of current density cannot be neglected in a comprehensive discussion on the retained dose and the possible bulk structure modifications induced in the sample. Measurements by a thermocouple are in agreement with the evaluated temperatures.[11]

Structure of the implanted layers

Conversion Electrons Mössbauer Scattering (CEMS) has been used to examine the structure of the implanted layers. Table 2 reports the relative areas of spectral components for samples implanted with current densities 50 and 200 $\mu A/cm^2$, whose spectra are shown in Figs. 3 and 4. Formation of iron-nitrogen compounds depends on starting structure and on current density. For both types of matrix the contribution of surface compounds is larger in samples implanted at 50 $\mu A/cm^2$ than in samples implanted at 200 $\mu A/cm^2$. This observed trend agrees with the values of the retained dose in the near surface region reported in Fig. 1.

Considering CEMS data of implanted quenched samples which in the virgin state correspond to martensite plus small amounts of retained austenite[7,8], the 50 $\mu A/cm^2$ current density implanted sample presents a contribution of surface compounds smaller than in the corresponding annealed sample, according to the measured retained dose. Moreover, in the sample implanted at 200 $\mu A/cm^2$ no ε-iron phase was detected. In

Table 1 - Microhardness values of the examined samples.

STEEL	CURRENT DENSITY ($\mu A/cm^2$)	IMPLANTED SIDE MICRO-HARDNESS (HV)	UNIMPLANTED SIDE MICRO-HARDNESS (HV)	EVALUATED TEMPERA-TURE (°C)
C85 ANNEALED (before implantation: 185 HV)	50	210±5	185±5	
	100	220±5	185±5	
	200	210±5	185±5	
C85 QUENCHED (before implantation: 960 HV)	50	600±10	520±10	320
	100	470±10	470±10	360
	200	410±10	400±10	450

Table 2 - Measured relative areas of Mössbauer components for samples implanted at 50 and 200 $\mu A/cm^2$ current density. Uncertainties are reported in brackets.

STEEL	CURRENT DENSITY	FERRITE	CEMENTITE	γ'-Fe_4N	ε-IRON PHASE			
					Fe-I	Fe-II	Fe-III	Fe-Q
C85 ANNEALED	50	30(5)	10(5)	5(5)			5(5)	50(3)
	200	72(5)	10(5)	10(5)		8(5)		
C85 QUENCHED	50	56(5)	7(5)	25(5)		12(5)		
	200	83(5)	10(6)	7(5)				

both cases there is a contribution of cementite and of γ'-nitride, meaning that the implantation in the considered range of current density produces considerable heating responsible for martensite tempering and for carbide precipitation. Partial depletion of interstitial carbon allows the formation of γ'-phase as occurs in pure iron[10] and in tempered and annealed steels[8,9].

It is interesting to compare our results particularly on the annealed implanted samples with those obtained by Longworth and Hartley[10] on nitrogen-implanted iron foils. They studied the effect of annealing after implantation and showed in particular that the implanted nitrogen atoms at the dose 4×10^{17} N^+/cm^2 rearranged in the CEMS investigated layer as follows: *i)* the relative area of ε-iron compnents increased by increasing the annealing temperature up to 325 °C and dropped to zero by annealing at 400 °C; *ii)* also the relative area of γ' component showed a maximum at 325 °C annealing temperature, going to zero by annealing at 500 °C; *iii)* the Mössbauer pattern of the sample annealed at 500 °C was corresponding to that of the iron-nitrogen martensite. Therefore they concluded that heating produces a change in the surface compounds composition and diffusion of nitrogen towards the bulk. The fact that the annealing at 500 °C destabilizes the nitrides formed by implantation on iron agrees with the low values of surface compounds relative areas encountered in our samples implanted at 200 $\mu A/cm^2$ current density and correspondingly heated at about 450 °C. The discrepancies between the relative areas of our surface compounds and those of ref. 10 are probably due to the different matrices employed, to the heating produced during and not after implantation and to the heating time.

Finally, it is to be noticed that in quenched implanted samples the predominant surface compound is γ'-Fe_4N. This means that the amount of nitrogen in the CEMS investigated layer is too poor to produce the interstitial richer ε phase, as a consequence of the implantation temperature that aids the nitrogen migration towards the highly defective bulk structure. The effect is obviously enhanced at the highest dose-rate. The easier migration of interstitials in a quenched structure is to be considered the reason for the measured dose in the quenched samples being lower than in the annealed. The fact that in quenched samples the retained dose remains practically constant above 100 $\mu A/cm^2$ current density may be explained by a balancing effect between the easier migration of interstitials in a defective structure and a progressive annihilation of defects by the tempering effect.

CONCLUSIONS

In the implanted quenched steel we have observed that: *i)* the measured dose is systematically lower than in the annealed steel; *ii)* the microhardness after implantation decreases by increasing the beam's current density and *iii)* carbide precipitation occurs jointly to a modest

formation of surface compounds.

The different behaviour with respect to the annealed steel may be accounted for by a tempering effect during implantation.

The behaviour of the measured dose versus the current density for the annealed steel and the correspondence between current density and temperature suggests that in this case the migration of the implanted nitrogen towards the bulk is essentially temperature-controlled.

ACKNOWLEDGMENTS

The authors are most grateful to Prof. P. Mazzoldi and Prof. I. Scotoni for stimulating comments in relation to this work.

Research sponsored in part by CNR, P.F. Metallurgia.

REFERENCES

1. See G. Dearnaley: Ion Implantation Metallurgy, ed. by C.M. Preece and J.K. Hirvonen (New York, N.Y. 1980) p. 1.
2. N.E.W. Hartley: Thin Solid Films 64, 177 (1979).
3. S. Lo Russo, P. Mazzoldi, I. Scotoni, C. Tosello and S. Tosto: Appl. Phys. Lett. 34 (10), 627 (1979).
4. S. Lo Russo, P. Mazzoldi, I. Scotoni, C. Tosello and S. Tosto: Appl. Phys. Lett. 36 (10), 822 (1980).
5. P.L. Bonora, M. Bassoli, G. Cerisola, P.L. De Anna, P. Mazzoldi, S. Lo Russo, I. Scotoni, C. Tosello and M. Maja: Mater. Chem. 4, 17 (1979).
6. P.L. Bonora, M. Bassoli, G. Cerisola, P.L. De Anna, S. Lo Russo, P. Mazzoldi, B. Tiveron, I. Scotoni, C. Tosello and A. Bernard: Nucl. Instr. Meth. 182/3, 1001 (1981).
7. G. Principi, P. Matteazzi, E. Ramous and G. Longworth: J. Mater. Sci. 15, 2665 (1980).
8. R. Frattini, G. Principi, S. Lo Russo, B. Tiveron and C. Tosello: J. Mater. Sci. 17, 1683 (1982).
9. R. Frattini, L. Giordano, S. Lo Russo, G. Principi and C. Tosello: Proc. Int. Conf. Appl. Mössbauer Effect, Jaipur (India) dec 1981. in press.
10. G. Longworth and N.E.W. Hartley: Thin Solid Films 18, 95 (1978).
11. L. Gratton, L. Guzman, A. Miotello, C. Tosello and G. Wolf: unpublished work.
12. N.E.W. Hartley: A.E.R.E. Report R-9065 (1978).

R P HOWSON

Sputter coating

SYNOPSIS

The reasons why sputtering and particularly d-c planar magnetron sputtering have emerged as preferred coating techniques for many applications is discussed. These are given as the simplicity and comparatively high rate of the system with a high degree of control and a large range of source material. It is shown that the system is suitable for the reactive preparation of many compounds and allows the use of low temperature substrates.

Examples of the commercial use of such systems is given.

THE AUTHOR

Department of Physics, University of Technology, Loughborough, LE11 3TU, Leicestershire

INTRODUCTION

Some time ago we started an investigation into the preparation of thin films onto large area substrates. These films had to be of a variety of metals and dielectrics. They had to be able to be produced at a high rate to achieve an economical cost, but had to be of the optimum quality with good control of thickness. The technique we chose after considerable investigation was that of sputtering.

This decision indicates the progress that has been made with a technique which had been relegated to a historical curiosity after forming the first method of making thin films in a vacuum. This revival has occurred because of the availability of high frequency power generators and the considered use of magnetic field to contain gas discharges.

In this paper we would like to indicate the features of the equipment that is available and why it becomes a preferred form of deposition for many processes. The resurgence of this technique has led to new areas of coating technology being developed and some of these will be described.

VACUUM PROCESSES

The residual atmosphere within an evaluated enclosure consists of atoms and molecules of volatile gas and some of the material being transferred within the chamber. The vital descriptive parameters of this situation are the mean free path between collisions of the constituents and the number of collisions of these with the chamber walls.

In the process of evaporation, a source is heated to liberate material which is transferred to a substrate with a low probability of collision; an important consideration in the formation of the film is then the flux of material onto the substrate compared to the number of collisions from the residual atmosphere.

In many situations the residual atmosphere dominates the process and if the gas is reactive then compound formation can result.

Estimates of the mean free path can be quickly made from the knowledge that it is linear with pressure and generally about the dimensions of laboratory apparatus at 10^{-4} mbar, i.e. 0.5 m in argon. However, the collision rate of the gas is $\sim 4 \times 10^{16}$ mol.cm^{-2}s^{-1} ($\sim 10^{-2}$ s.per monolayer) at this pressure. A high rate of deposition is typically about 30 nm/sec, about the same condensation rate as that of the gas, and this fact must be appreciated when the structure of the film is considered.

The movement of particles within a vacuum can be influenced by electric and magnetic fields. Of course only charged particles can be influenced in this way, but where operation occurs at pressures which result in mean free paths being considerably less than the dimensions of the apparatus then a considerable "knock-on" of neutral particles occurs. Considerable energies can be acquired by charged particles from electric fields, much greater than those that are given by thermal processes. A tungsten source evaporating at say 3,000°C will give particles with an average energy of around 0.25 eV, which is equivalent to a charge particle being accelerated through a potential difference of a quarter of a volt. Voltages of several hundreds of volts are typically applied to gas discharges.

In a vacuum process we are generally only concerned with the so -called "abnormal" gas discharge which is characterised by a current density which rises rapidly with the applied voltage, a uniform current density from the electrodes and a dark region around the cathode in a d-c discharge, which is given the name of the Crooke's, or cathode, dark space.[1] This is attributed to the distance required by electrons leaving the cathode to acquire the energy necessary to ionise the residual gas. When the gas is ionised it becomes effectively fully conducting in that all the potential applied appears across this dark space which has dimensions of between 10 and 20 "mean free path" lengths of electrons in the gas, this is generally taken as 14. The process is then one of positive ions being accelerated across the dark space to be driven into the cathode at high energy liberating material and electrons to form the sputtered material and electron source to maintain the discharge. The fact that the sputtered material must suffer many collisions on moving from the region of the cathode target make the process very inefficient. Also the substrate is around ground potential (the whole of the luminous region of the plasma, the positive column, is at the potential of the anode, which is generally ground) so that it will be subject to collisions from the electrons in

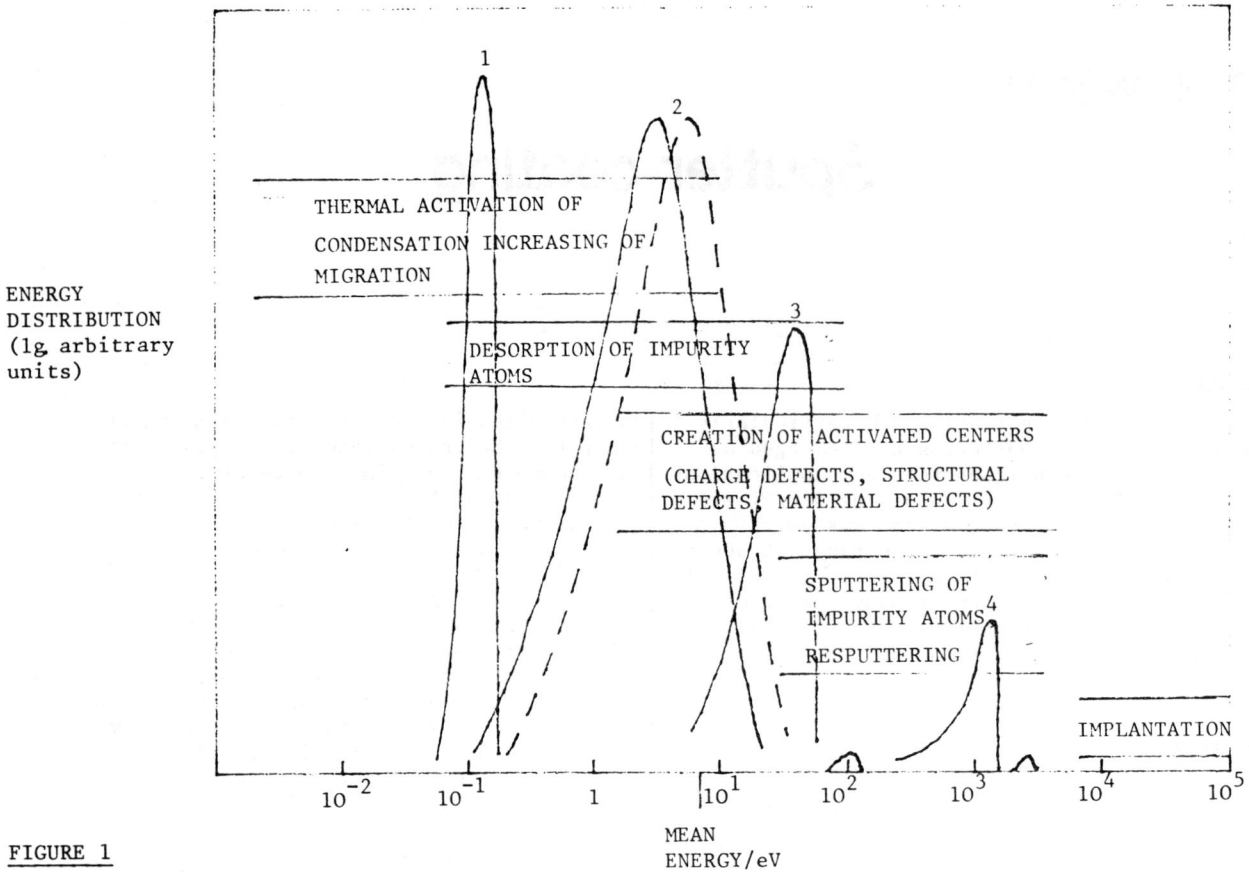

ENERGY
DISTRIBUTION
(1g arbitrary
units)

THERMAL ACTIVATION OF
CONDENSATION INCREASING OF
MIGRATION

DESORPTION OF IMPURITY
ATOMS

CREATION OF ACTIVATED CENTERS
(CHARGE DEFECTS, STRUCTURAL
DEFECTS, MATERIAL DEFECTS)

SPUTTERING OF
IMPURITY ATOMS
RESPUTTERING

IMPLANTATION

MEAN
ENERGY/eV

FIGURE 1

The energy distribution and effects of ions
in vacuum processes.

1. the thermal energy of vapour particles.

2. the distribution of sputtered particles
    at the target and at the substrate
    (dotted line)

3. Ion and neutral particle energies in
    bias sputtering, and

4. in ion plating for the same mean energy.

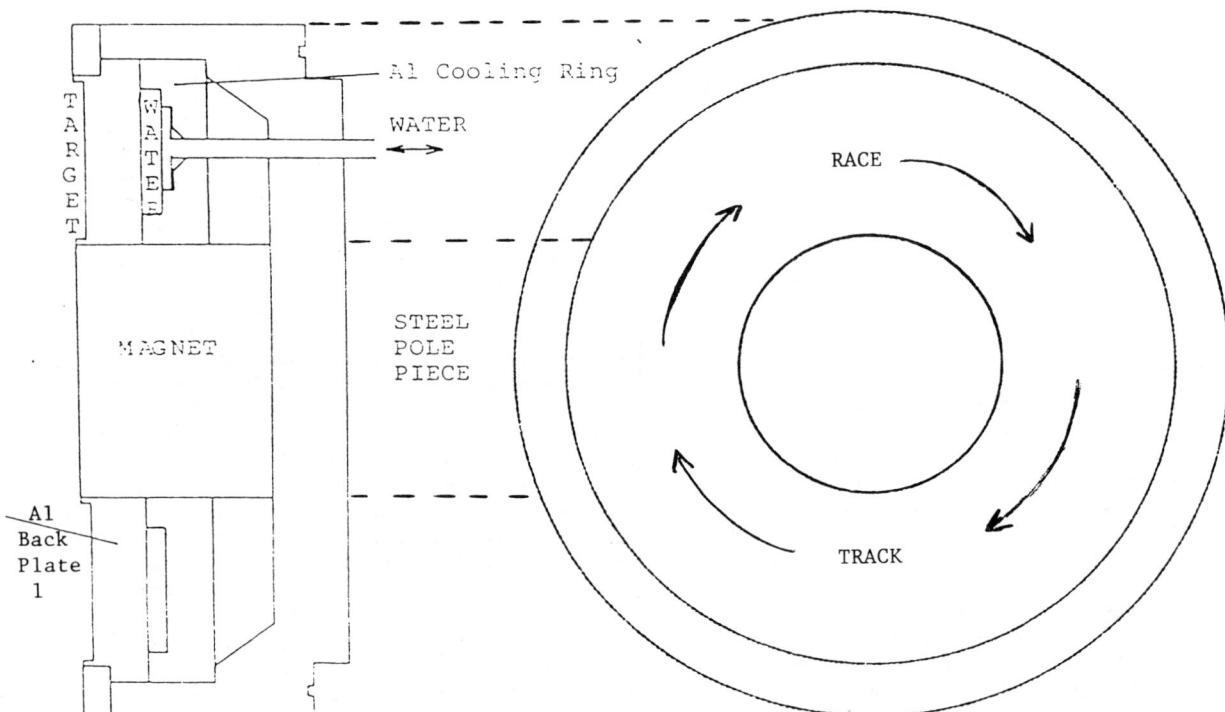

FIGURE 2

A circular planar magnetron.

the discharge, causing radiation damage and considerable heating effects. It is known that the deposition of energetic species can lead to purer, better adhered coatings which often have a better structure[2] The influence of the energies of bombardment is indicated in Figure 1[3] which shows that apart from the bombardment of material to cause re-sputtering, some bombardment of the growing film can be very advantageous.

FILM FORMATION

The most obvious way to make a thin film in a vacuum is to heat a material source and allow the source to evaporate through as near a perfect vacuum as possible to condense on a clean surface. This is realised in molecular beam epitaxy when semi-conductor material is condensed onto single crystal surfaces, but is not a realistic process for large area coatings. Conventional evaporation is done at a much higher pressure; purity can sometimes be maintained by using very high rates of condensation from electron beam heated sources, but the condensate is generally of low energy which leads to uncertain adherence of the coating to the substrate. Ways of adding energy to the condensing atoms have been devised using ionisation of the material with an electron beam and subsequent acceleration (Bunshah)[4] by having the material evaporate through a gas discharge directed towards the substrate which has been called ion-plating (Mattox).[5] (Re-sputtering can be avoided using clusters of atoms of the evaporating species (Takagi),[6] by adding energy through a separate beam of neutral or ionised inert particles (Weissmantle)[7] and by using a plasma gun to provide energetic ionised reactive gas (Ebert).[8]

Conventional sputtering provides an energetic condensate which, unfortunately, is dissipated through gas phase collisions; these must be reduced. Operation at lower pressure is the answer, but unfortunately, this results in being unable to maintain an electric abnormal discharge with the straightforward application of a potential. It is the achievement of intense gas discharges at low pressures which has led to the resurgence of sputtering as a technique of providing high quality films at a high rate.

In general there are three ways of achieving this: One, use a separate source of elect-rons to maintain the discharge. This is generally achieved using a high electron current from a hot filament to maintain a separate d-c discharge. Secondary discharges in higher pressure chambers may also be used. Two, use a high frequency electric field, 50 kHz or greater, when electrons are not lost from the discharge, but oscillates within it to allow much lower pressure of operation.[2] Generally an industrial frequency of 13.56 MHz is used. This process has the additional advantage of being able to sputter insulating material because of the alternative bombardment of the target with ions and then electrons which prevent charge build-up. It is, however, much more expensive and inefficient than a d-c process. Three, contain and increase the path of electrons within the discharge region using a magnetic field. This proved to be successful when an external magnetic field was applied (Thornton)[9] but has become much more powerful with the use of re-entrant magnetic fields originating behind the cathode target, the planar magnetron.[10] The principal of this technique is the helical path taken by electrons in a crossed electric and magnetic fields giving much longer effective paths in the ionization region of the discharge allowing longer "linear" mean free paths to be used and, hence, much lower pressures. The problem with external fields was that the electrons were eventually lost at the open ends of the magnetic 'flask'. In the planar magnetron, the magnetic field comes through the cathode, travels effectively parallel to the surface and goes back through the target. The movement of the electrons that occurs mutually perpendicularly to the electric and magnetic fields due to Hall effect forces, can be made to allow the electrons to circle within a continuous 'race-track' created by the geometry of the permanent magnetic field. This is gener-ally either a circle as illustrated in Figure 2, or a flattened oval which can be as long as required; 3 metre lengths have been made. The pressure of operation is determined amongst other factors by the minimum radius of curvature designed into the race-track and the strength of the magnetic field. By this simple modification to a sputtering target, the pressure of operation is reduced a hundred-fold allowing much more efficient transfer of material to the substrate, the voltage of operation is reduced by ten-fold and, what emerges as a very important factor, the bombardment of the substrate by the electrons in the discharge to be prevented, allowing the coating of sensitive substrates at low tempera-tures without introducing radiation damage.

A further advantage of an energetic conden-sate is that reactions can be encouraged on the substrate surface to give compound formation. Oxides are the most popular choice with the use of a gas mixture of argon, with it's high sputtering yield for metals, and oxygen to form the oxide with the growing film, but nitrides can be made from activated nitrogen or ammonia, and carbides from a hydro-carbon gas such as methane. There appears to be no real restric-tion on the process because of the very high energies that are available; what is required is an appropriate volatile gas. We have added fluorine to a growing film from fluon and also added one metal to another to form a mixed metal oxide.[11] Indium was sputtered in an atmosphere of tin tetramethyl, oxygen and argon to give indium tin oxide films.[12]

SPUTTERING SYSTEMS

The design of sputtering sources, the rates obtained and the difficulties experienced with simple element sputtering and the further ones created with reactive sputtering have been comprehensively reviewed by Vossen[2] and Schiller[3] Of particular interest is the control of properties that can be achieved with a reactive process which is illustrated in Figure 3 where it is shown that the stoichiometry of a highly important optical material, TiO_2, can be controlled through the deposition conditions.[20]

We have extended this system with the addition of a r.f. discharge directed towards the substrate to control the energy put into the depositing film,[13] the apparatus to do this is illustrated in Figure 4.

We have sought to create a technique which will create conducting oxide films onto room temperature substrates at a high rate and in a continuous process from the simplest starting materials. We chose a fully reactive process with a metal source and a substrate which is a roll of plastic wound continuously from roll-to-roll during the deposition process.

The precision control of rate, combined with a high sputtering rate, capability, and operation at a relatively high pressure to provide the reactive gas, has led us to the

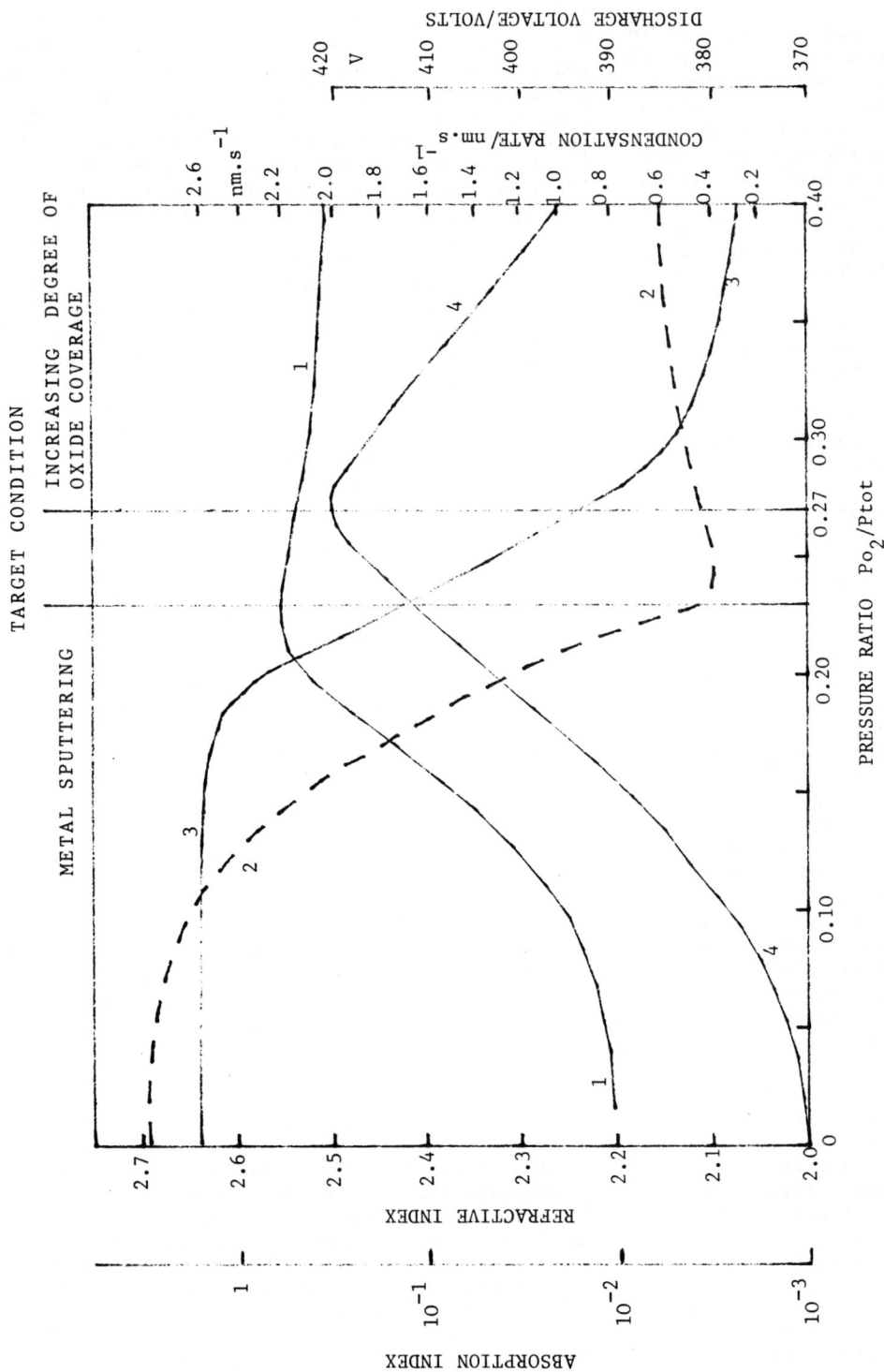

FIGURE 3

The optical properties of Titanium oxide (20)
as a function of the deposition conditions.

1. Refractive index,
2. Absorption index,
3. Condensation rate, and
4. Discharge voltage.

system of d-c planar magnetron sputtering of the metal or alloy in an atmosphere of argon and oxygen. The ion bombardment of the substrate provided by a radio frequency discharge activates the oxygen reacting on the substrate surface to give higher rates of deposition. In order to achieve optimum properties in the films we have found it necessary to control the power to the sputtering source and the oxygen gas admission rate to within 0.1%. The conditions were adjusted in accordance with the properties of the film continuously monitored in the roll coating apparatus. We have elucidated the advantages of such a system as :

1. A high rate of sputtering is obtained because of the confinement of the plasma close to the target surface. The limit is determined by a factor involving the thermal conductivity of the target, the efficiency of the water cooling, the melting point of the sputtering material and its sputtering yield.

2. The heating of the substrate by stray electrons is low. This is the limiting factor for high rate conventional sputtering in many cases.

3. Sputtering can take place at a low gas pressure because of the increased path length of electrons within the plasma and prevention of their escape. This is achieved by the crossed electric and magnetic fields.

When this sputtering source is used in the presence of a reactive gas to create a compound, further advantages ensue:

1. The rate of deposition can easily and accurately be controlled to allow balanced processes to be maintained. In particular, the high surface ion current density allows sputtering of the metal from the metal target even in a reactive gas, with the compound formation taking place on the substrate surface rather than on the target surface.

2. The gas pressures required for sputtering and for producing reactive processes are compatible.

3. Alloy metal targets can easily be sputtered to give mixed metal oxides of similar metal ratios.

If an additional field is applied to the substrate surface, a discharge can be created which bombards the growing film surface with ions. The advantages of 'ion plating', used in conjunction with sputtering, are:

1. The reactive gas is activated by ionization and driven towards the substrate surface by the ion plating radio frequency field.

2. The addition of surface energy by the reactive and neutral gas atoms to the growing film results in control of structure and nucleation phenomena without using high temperature substrates.

3. The gas pressures needed for the sputtering and ion plating discharges are compatible.

A roll coating system offers advantages for the development of coatings, especially with the reactive process. These are:

1. High rate equilibrium processes are possible. Long substrate lengths allow lead-in times which eliminate shuttering disruption.

2. Rapid preparation gives pure deposits. The ratio of depositing species to residual impurity gas is high.

3. High experimental productivity is obtained; many samples can be produced on one roll in one evacuation. In-situ monitoring facilitates parameter optimisation in a short time.

4. The results are applicable to commercial processes.

The rate of sputtering was over 0.5 μm $min^{-1}kW^{-1}$ and the power used was about 3.5 kW into the one target.

The properties obtained with indium oxide films are shown in Figure 5 where the degree of control of the properties is clearly indicated[14] This is only attained because of the very high stability of the planar magnetron process.

COMMERCIAL SYSTEMS

The ability of the magnetron source to be a high rate source of a complex alloy, allowing transfer of the material to a substrate at a high rate whilst maintaining composition proportions, has led to this process being adapted by the semi-conductor industry for the coating of Al-Si alloys as a conducting layer on the surface of a single crystal silicon intended for integrated circuits.

Generally a load lock system is used with the target operating continuously to avoid surface oxidation of the aluminium that occurs when the system is taken up to atmosphere. It is difficult to remove oxides from the surface of metals because of their very low sputtering rates and high secondary electron emission coefficients which give a low ion bombardment for a given current.

The coating of plate glass with films for both solar control and for heat reflection has been reported by vacuum plant manufacturers.[15, 16] This is done with planar magnetrons of up to 3 metres long operating at high rate with input powers of up to 80 kW into one target. Equipment is available to prepare films of TiN where a very short cathode-substrate distance is used to create a very high energy on the substrate surface to stimulate the reactive process. When reaction is required to be avoided, for example in the preparation of very thin films of silver, very strong magnetic fields can be used to keep energy away from the substrate. Weakening of the magnetic field puts the energetic part of the plasma onto the substrate surface to encourage oxide reactions.[16]

The coating of plastic sheet in a large machine has been reported and the technique has been favourably critically reviewed for the preparation of dielectric interference coatings.

A variation of the planar magnetron using magnets within a post have been reported by General Engineering[17] and Ford.[18] The former system uses oscillating magnets within the post to give a race-track around the post; the latter spins the magnets within it, using a race-track up and down the post. Both these systems give good utilisation of the material of the post cathode to coat piece parts distributed around the periphery of the vacuum chamber.

Point sources of magnetically contained plasma have been used industrially for large scale coating and are available for coating under the name of 'S' guns. [19]

CONCLUSION

The use of radio frequency discharges and magnetic containment of d-c discharges has led to sputtering systems being created which have advantages over some of the evaporation sources that were used hitherto, and creates new possibilities for the creation of coatings. The principal advantage lies in the quality of the coating that is produced and the ability to produce this uniformly over large widths of substrate that are moved under the source.

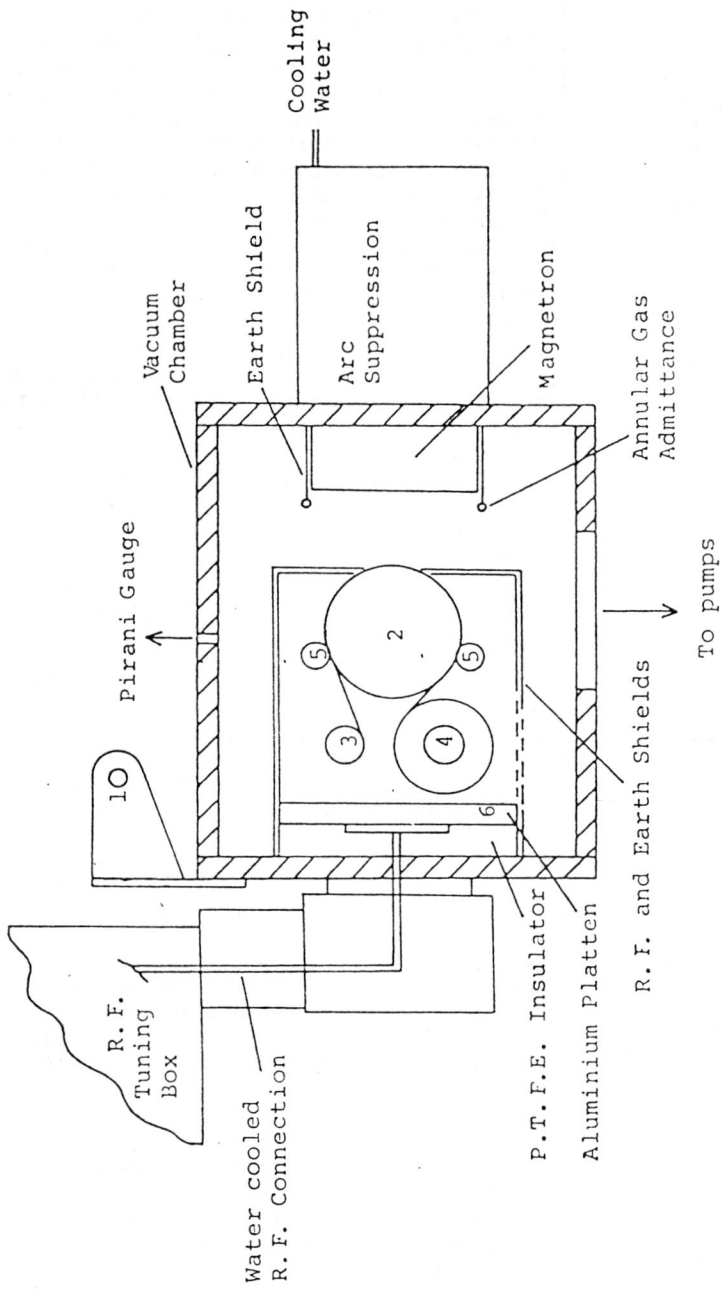

FIGURE 4

Apparatus for the coating of sheet plastic transferred from roll to roll inside a vacuum using a planar magnetron source and radio frequency ion plating.

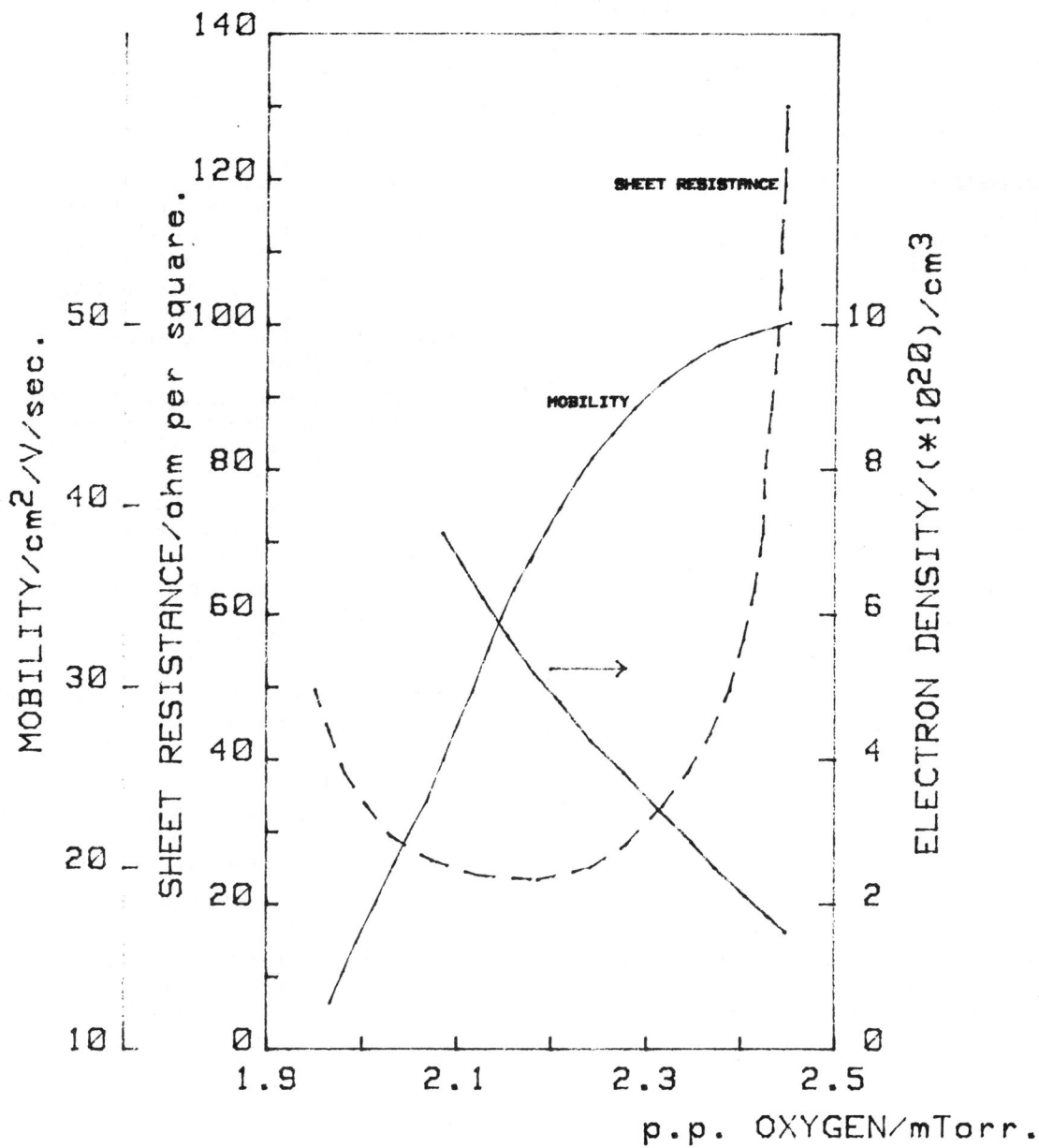

FIGURE 5

The Hall electrical conduction parameters for a reactively sputtered film of indium oxide as a function of the deposition conditions. There was a partial pressure of argon of 3,5 mT and the magnetron current was 8 amps at a potential of 400 volts.

The rate of deposition is high, not as high as electron beam evaporation, but offers a tremendous range of materials; tungsten can just as easily be deposited as copper. The high energy of deposition allows reactive processes to be controlled on low temperature substrates with a very high degree of precision.

Problems remain, the poor power efficiency and target material utilisation and difficulties in the fabrication of a sheet target, but the exciting possibilities of the creation of exotic refractory metals and compounds onto almost any substrate, more than compensate for these.

ACKNOWLEDGMENTS

The author is pleased to acknowledge the considerable help he has received from M.I. Ridge and C.A. Bishop.

REFERENCES

1. Chapman B. Glow Discharge Processes, Wiley, 1980.

2. Vossen J.L. & Kern W. Thin Film Processes. Academic Press 1978.

3. Schiller S, Heisig U. & Goedicke K. Ion Deposition Techniques for Industrial Applications. Proc. 7th Int. Vacuum Congress & 3rd Int. Conf. Solid Surfaces, 1545-1552, Vienna 1977.

4. Bunshah R.F. Production of Thick Films by PVD Techniques and their Applications. Ibid 1553-1558.

5. Mattox D.M. Mechanisms of Ion Plating. Proc. I.P.A.T. 11-10, London 1977.

6. Takagi T. et al. The Vapourized-Metal Cluster Ion Source and its Application to film formation. Proc. 7th Int. Vac. Congress & 3rd Int. Conf. Solid Surfaces 1603-1606, Vienna 1977.

7. Gautherin G, Schwebel C & Weissmantle C. Improved Ion Beam Sputtering and its Application. Proc. 7th Int. Vac. Congress & 3rd Int. Conf. Solid Surfaces, 1579-1582, Vienna 1977.

8. Ebert J. Activated Reactive Evaporation. Proc. SPIE, 325, 29-38 (1982).

9. Thornton J.A. & Penfold A.D. Cylindrical Magnetron Sputtering. As Ref. 2 pp 76-113.

10. Waits R.K. Planar Magnetron Sputtering. As Ref. 2 pp 131-173.

11. Avaritsiotis J.N. & Howson R.P. Composition and conductivity of fluorine-doped conducting indium oxide films prepared by reactive ion plating. Thin Solid Films 351-357, 77 (1981).

12. Turner P, Howson R.P. & Bishop C.A. Optical Thin Films obtained by Plasma-induced CVD. Thin Solid Films 253-258, 83, (1981).

13. Ridge, M.I., Howson R.P., Avaritsiotis J.N. and Bishop, C.A. Reactive ion plating with a Planar Magnetron Sputtering Source. As Ref. 5 pp 21-26.

14. Ridge M.I. & Howson R.P. Composition control in conducting oxide thin films. Thin Solid Films 121-127, 96, (1982).

15. Grubb A.D., Mosakowski T.S. & Overacker W.G. Production Techniques for high volume sputtered films. Proc. SPIE 325, 74-81 (1982).

16. Munz W.O. & Reineck S.R. Performance and sputtering criteria of modern architectural glass coatings. Ibid, pp 65-73.

17. General Engineering Co. (Radcliffe) Ltd.

18. Hoffman D.W. Design and capabilities of a novel cylindrical post magnetron sputtering source. Thin Solid Films - to be published.

19. Fraser D.B. The Sputter and S-Gun Magnetrons. As ref. 2, 131-173.

20. Schiller S, Beister G, Schneider S, Sieber W. Features of an in-situ measurements on absorbing TiO_2 films produced by reactive d-c magnetron-plasmatron sputtering. Thin Solid Films 475-483, 72 (1980).

J P COAD and J E RESTALL

Sputter-ion plating of coatings for the protection of gas turbine blades against high temperature oxidation and corrosion

SYNOPSIS

Considerable effort is being devoted to the development of overlay coatings for protecting critical components such as turbine blades against high temperature oxidation, corrosion and erosion damage in service. The most commercially advanced methods for depositing coatings are electron beam evaporation and plasma spraying.

Sputter Ion Plating (SIP) offers a potentially cheaper and simpler alternative method for depositing overlays. Experimental work on SIP of CoCrAlY and NiCrAlTi alloy coatings is described. Results are presented of metallographic assessment of these coatings, and of the results obtained from high velocity testing using a gas turbine simulator rig.

THE AUTHORS

Dr Coad is in the Materials Development Division, AERE Harwell, Oxfordshire; Dr Restall is with the National Gas Turbine Establishment, Pyestock, Farnborough, Hampshire.

INTRODUCTION

Protective coatings are routinely employed in gas turbine engines in order to inhibit premature failure of high temperature components by oxidation, corrosion and erosion processes in service. The aerofoil section in particular of components such as turbine rotor blades and nozzle guide vanes manufactured in nickel- and cobalt-base superalloys requires protection against the hot gas environment in which it has to operate. Environmental aspects are complex and depending on the application for which the engine is used, include a high velocity gas-stream which, depending on the combustor design, purity of the fuel etc, may range from being simply oxidising to corrosive, erosive or even locally reducing.

The low levels of contaminants such as sulphur and vanadium permitted in the high quality distillate fuels which are used in aircraft gas turbines together with fairly simple diffusion-type aluminide coatings minimise oxidation and corrosion damage in service. An example of a turbine rotor blade withdrawn during routine overhaul of a civil aero-engine after several thousand hours service is shown in Figure 1 (a). The blade, which was manufactured in a wrought nickel superalloy and subsequently pack-aluminised, showed no significant corrosion damage although some oxidation attack of the coating was evident. Failure of the coating would have occurred at much longer time due to spallation of the protective alumina film during thermal cycling and depletion of aluminium through interdiffusion between the coating and substrate to a level where a stable protective Al_2O_3 scale could not be maintained.

In contrast to the situation for high flying aircraft, gas turbine engines that are required to operate close to the sea, e.g. for maritime patrol or for ship propulsion, ingest significant quantities of sea-salt spray into the engine intake. The salt particles subsequently become entrained in the hot combustion gas stream and are deposited at the surfaces of the turbine blades where they react to produce accelerated hot-corrosion[1-10]. An example of a severely corroded turbine rotor blade withdrawn after approximately 1500 hours service from a light-weight marine gas turbine engine is shown in Figure 1 (b). Improved blade coatings are required both to combat hot corrosion and to improve the resistance to oxidation of newer blade materials which are having to operate at higher and higher temperatures.

Some details of coating materials, their associated process technology and results of engine and corrosion rig experience on turbine aerofoil components may be found elsewhere[11-16]. Most of the coating compositions in general use for protecting turbine blades and vanes against high temperature oxidation and corrosion processes are based essentially on their ability to form protective scales based on alumina or chromia. During the past thirty years pack-cementation has

1. Turbine rotor blades after service. (a) after
   several thousand hours in an aero-engine,
   (b) after 1500 hours in a marine turbine.

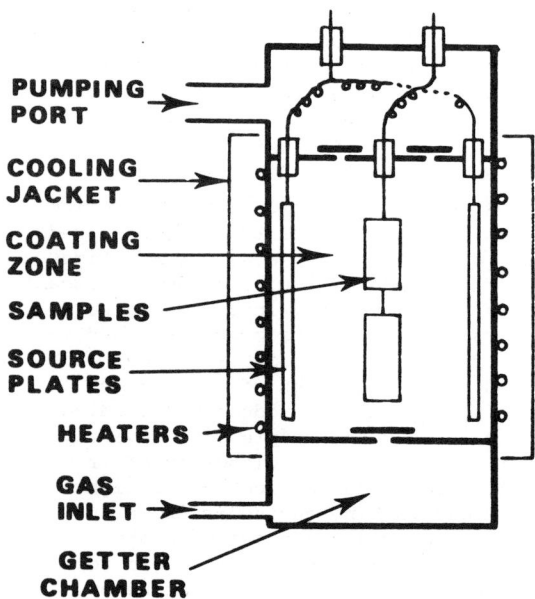

PUMPING
PORT

COOLING
JACKET

COATING
ZONE

SAMPLES

SOURCE
PLATES

HEATERS

GAS
INLET

GETTER
CHAMBER

2. A schematic view of the Sputter Ion Plating
   (SIP) system.

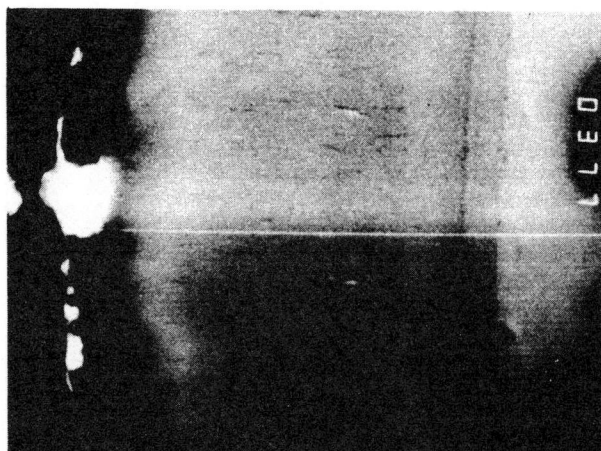

3. A secondary electron image of part of a
   CoCrAlY-coated nickel alloy turbine blade.

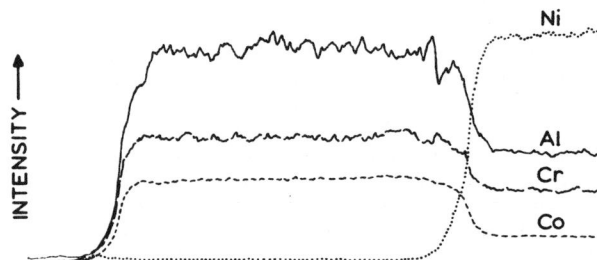

4. Intensity profiles for the elements Ni, Al,
   Cr and Co across the CoCrAlY coating shown
   in Figure 3.

13.2

been the most widely used commercial process for depositing aluminide or chromium-enriched coatings on superalloys. However, since the mid-1970s these diffusion-type coatings have been challenged by developments in so-called overlays which offer a superior combination of ductility and oxidation/corrosion resistance. A wide range of overlay coating compositions is available, the most commercially important of which are based on MCrAlY alloy (where M is Ni, Co or Fe). A number of methods for applying overlay coats are industrially available or at an advanced stage of development. The most important of these include the commercial electron beam evaporation process and plasma spraying. However, these techniques when used for coating blade aero-foils require a relatively large capital investment and interest is being directed towards simpler and cheaper alternatives.

One such alternative is Sputter Ion Plating (SIP) which offers numerous advantages including good throwing power, combined with excellent adhesion, good control of deposit morphology, flexibility in choice of coating composition, and ease of coating components of complex geometry without the need to manipulate the samples during coating operations. In this paper, experimental work is described for SIP of overlay coatings based on CoCrAlY and NiCrAlTi compositions. Results obtained from laboratory oxidation and high velocity corrosion testing using a gas turbine simulator rig are presented.

EXPERIMENTAL METHOD

SIP has been described in detail elsewhere (18, 19) but an outline of the technique is given here for clarity. As the name indicates, the method uses sputtering from solid targets to produce the flux of coating materials (mostly atoms) and an electrical bias is applied to the samples to give an ion polishing effect as described by Mattox and MacDonald (20). The system is shown in Figure 2. In the usual geometry plates of the coating material are placed round the vertical walls of the stainless steel coating chamber and turbine blades or other samples are suspended in the centre of the chamber. Although the source plates are insulated from the chamber walls, their proximity may facilitate control of the source temperature if required. For high temperature oxidation resistant coatings, however, the deposition temperature is usually arranged to be high by a combination of relatively high sputtering power density at the source plates and restricted cooling of the chamber wall.

The sequence of operations is as follows: a flow of extremely pure argon at reduced pressure (typically 10 to 100 mtorr) is established through the chamber and the samples and the source material are then preheated to about 300°C. Following preheating, the samples may be cleaned by ion bombardment, if desired, by applying a large negative voltage to them, thus establishing a glow discharge in the chamber from which ions are attracted towards the sample surface. During coating a large negative voltage, e.g. 1kV, is instead applied to the source plates, and the incident ions sputter material from the surface. This material (principally atoms) moves randomly through the coating zone with a short mean free path until it encounters a surface. The proportion of material from the source which reaches the turbine blades forms the coating, and the nascent deposit is "ion polished" by low energy ions attracted to the surface by a small applied negative voltage (typically 10 to 100V).

Coating rates in SIP are not large (up to 10 μm h⁻¹) but adequate thicknesses of high temperature oxidation resistant alloys (typically 100 to 125 μm) can be deposited in an overnight run onto a batch of blades. The technique is simple, requiring only a two-stage rotary pump and no sample manipulation, since the throwing power is excellent due to the extended sources and short mean free path.

A number of variations to the technique have been examined during this work. The first to be tried was the application of an axial magnetic field by analogy with magnetron sputtering. The field was effected by winding a coil round the cooling jacket and applying a DC voltage. At magnetic fields up to 100 gauss, the coating rate could be increased by about one-half, or the deposition temperature reduced for a given coating rate; the differences are only modest due to the relatively high pressure used in SIP. Since the economic use of SIP relies on large numbers of samples being processed at each loading, changes in the coating geometry were studied. A difficulty with scale-up of the simple geometry shown in Fig. 2 is that as the vessel diameter is increased, the coating volume to source area ratio and the source to sample distances (average) increase, so that discharge conditions and coating rates change. A way of more accurately predicting the conditions in a large unit is to change to a parallel plate geometry. Instead of placing source plates round the wall, the material is formed into two large parallel sheets (vertical in this equipment) separated by a distance large enough to accommodate rows of samples. Once coating conditions are optimised in this configuration, the linear dimensions of the sheets may be increased at will and only the discharge current will be altered (increasing in proportion to the source area) provided the sheet separation is held constant. Many of the experiments to be described were carried out using the parallel plate configuration, and results were indistinguishable from those performed using the cylindrical geometry.

An advantage of sputtering techniques is the ease with which alloys can be deposited, provided significant bulk diffusion cannot occur within the source material. Thus, any alloy may be deposited as a coating from source plates manufactured from the same alloy, since a dynamic equilibrium is rapidly established wherein supply of fresh material to the surface for sputtering is of the bulk composition, provided the plate composition is homogeneous. In practice, such alloys often consist of a number of different phases which because of microsegregation effects may lead to slight discrepancies in analysis between coating and source. In this work, such discrepancies have always been less than the possible errors in the analytical techniques and also less than the variation allowed in the specification of the source composition. A further advantage of sputtering is that the composition of the deposit may be varied by adding appropriate materials to the source plates. The materials used for the experimental work outlined in this paper have been based on selected compositions in the CoCrAlY range and experimental alloys in the NiCrAlTi system identified at NGTE as having good resistance to high temperature oxidation and corrosion. The two CoCrAlY alloys receiving most attention had the following nominal compositions (wt %):-Co-20%Cr-9%Al-0.3%Y and Co-25%Cr-9%Al-0.3%Y. The nominal compositions of the NiCrAlTi alloy plate sources used were: Ni-37%Cr-3% Al-2%Ti and Ni-35%Cr-8%Al-3%Ti respectively. The choice of these

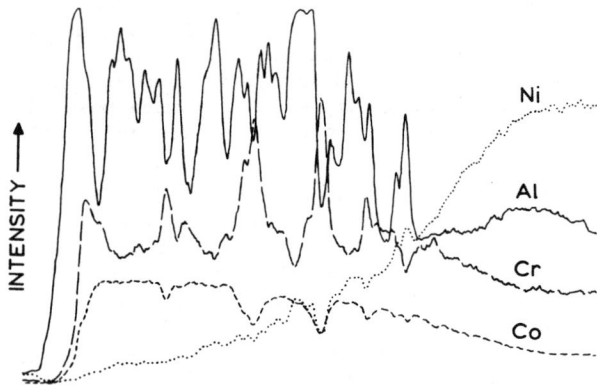

5. Intensity profiles for the elements Ni, Al, Cr and Co across a CoCrAlY coating exposed to air at 1100°C for 4 hours.

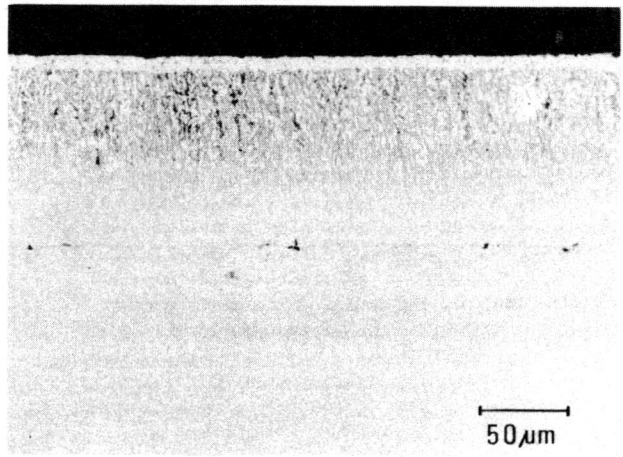

8. An optical micrograph of a CoCrAlY coating after 4 hours in air at 1100°C.

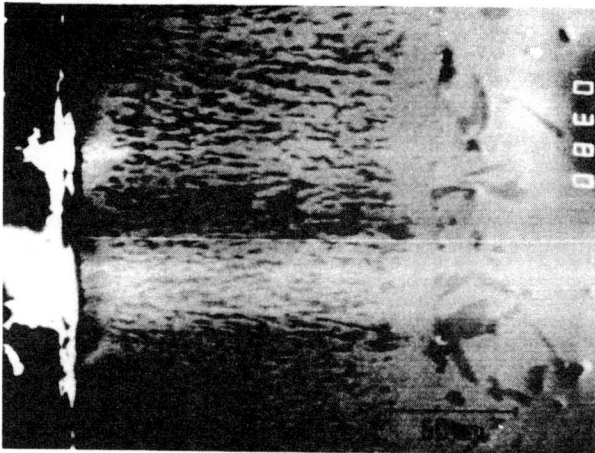

6. A secondary electron image of a CoCrAlY coating after oxidation. The image contains the line analysed in Figure 5.

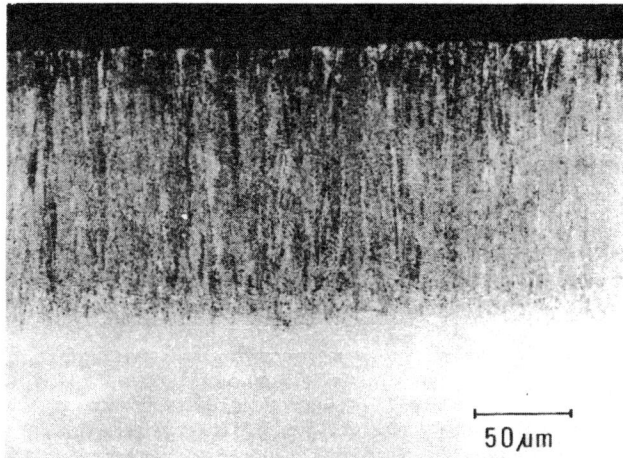

9. An optical micrograph of part of a NiCrAlTi-coated nickel alloy turbine blade.

7. An optical micrograph of part of a CoCrAlY-coated nickel alloy turbine blade.

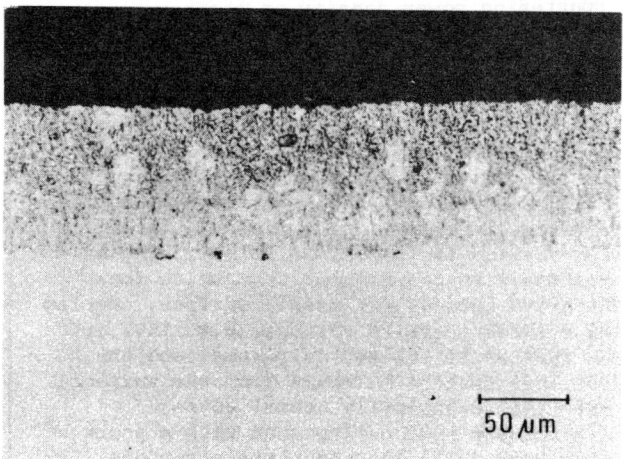

10. An optical micrograph of a NiCrAlTi- coating after 4 hours in air at 1100°C.

coating alloys was based on corrosion rig and marine gas turbine evaluation of bulk alloys which has shown that a high chromium content enhances resistance to aggressive corrosion by sea-salt, whilst aluminium enhances the oxidation resistance of superalloys.

Also in this work Y and/or Al strips have been added to standard MCrAlY plates in order to study the effect of increases in the concentrations of these elements. The analyses of the new coatings so formed must be determined empirically, since they cannot be accurately predicted from relative sputtering yields. Nevertheless, this provides a simple method of studying a range of compositions without the need for fabricating fresh targets. If more controlled work is needed on a specific composition, then sheets with that analysis can be formed or composite targets with sheets of different elements or alloys made. A variety of components have been used for the experimental work ranging from different designs of turbine rotor blades employed for aero and marine engines to very simple-shaped laboratory testpieces.

EXPERIMENTAL RESULTS

The flux of coating atoms approaching an isolated sample placed on the axis of the cylinder formed by the source plates has cylindrical symmetry. There is therefore a tendency for more coating material to arrive at the leading and trailing edges of a blade aerofoil than on the pressure and suction surfaces. There is also a greater flux at the suction (convex) surface than at the pressure (concave) surface. The resulting variation in flux as measured by the arrival rate at the surface may be about a factor of two. In ion plating, however, the flux of ions attracted to the (biassed) surface varies in a similar (or even greater) manner, and these ions tend to re-sputter material from the surface. Thus, the coating thickness distribution results from two processes which are of opposite effect and, as a result, variations are not as large as might be expected from consideration of coating flux alone. We have found that there is a limit to the amount of "ion polishing" that can be permitted, however, since at bias voltages high enough to give an almost uniform coating thickness there is an unacceptable variation in coating composition round the aerofoil, and this is more important than maintaining uniformity in thickness.

Standard metallurgical preparation procedures were adopted for assessing the quality and thickness variation of coated sections machined from the blade aerofoils. The etchant used for CoCrAlY coatings was 1% hydrofluoric acid in a mixture of 1 part nitric acid, 1 part hydrochloric acid and two parts water. NiCrAlTi coatings were etched in a solution of 6 parts acetic, 6 parts lactic, 2 parts hydrochloric and 1 part nitric acid. In each case, the etchant darkens the aluminium rich phases in the microstructure and the staining pattern produced gives a good idea of the uniformity (or otherwise) of composition around the aerofoil. More detailed analyses of the coating are obtained using the electron microprobe. The most useful forms of EPMA output have been found to be X-Ray line profiles from within the substrate through the coating to the surface, and small area analysis. Each type of datum is recorded from different parts of the aerofoil section being analysed. Figure 3 is a secondary electron image of part of a CoCrAlY coating at the leading edge of a nickel alloy blade: the sample is in the as-coated condition. A white line can be seen

drawn across the centre of the image and this is the line along which the X-Ray distribution scans for different elements shown in Figure 4 were obtained. In order to obtain the plots in Figure 4 the X-Ray detector is set on a particular element and the peak intensity is plotted as the electron beam traverses the line visible in Figure 3. The detector is then set for the next element of interest and the process repeated. For all the coatings, it was observed that there is little variation in intensity within the coating or substrate regions, although the coating/substrate interface is clearly delineated.

In contrast to Figure 4 a series of line scans is shown in Figure 5 for a similar CoCrAlY coating after four hours exposure in air at 1100°C. Various features visible in the Figure include a protective film of Al_2O_3 formed at the outer surface of the CoCrAlY coating. A zone depleted in aluminium exists between the thin film of alumina and the bulk of the coating which has a duplex microstructure. The latter overlays a diffusion band formed between the overlay coating and substrate alloy during high temperature exposure. After a further oxidation period of 24 hours at 1100°C only slight changes occurred within the microstructure which were associated with coarsening of the phases present in the original coating. A secondary electron image of the area analysed in Figure 5 is shown in Figure 6. Optical micrographs of coatings deposited on areas of the same blade aerofoils as those in Figures 3 and 6 are shown in Figure 7 and Figure 8 (respectively).

Optical micrographs of NiCrAlTi coatings on turbine blades are shown in Figure 9 and Figure 10. The coating in Figure 9 is in the as-deposited condition whilst the thinner coating in Figure 10 had been heated in air at 1100°C for four hours. This heat treatment also produced a distinct two-phase structure in the coating.

In addition to the fairly simple laboratory oxidation tests some high velocity corrosion rig tests were carried out on selected coatings at the Admiralty Materials Technology Establishment at Holton Heath. In these, gas turbine engine simulator rig coated testpieces were subjected to a high velocity gas stream produced by combustion of Dieso fuel (1% S maximum) with 1 ppm of seawater added as an aqueous solution. Tests were carried out for durations up to 1200 hours at 750°C under these very corrosive conditions. At the end of the tests the specimens were subjected to detailed metallurgical assessment. SIP NiCrAlTi coatings approximately 80 μm thick provided excellent corrosion resistance to the nickel alloy substrate even after 1200 hours exposure whereas aluminide and other coatings were completely degraded after only 100 hours exposure for the same conditions.

CONCLUDING REMARKS

A relatively simple laboratory-scale SIP apparatus has been used to deposit overlay coatings on the surfaces of turbine rotor blades of the type used in advanced engines, and also for coating of laboratory testpieces. The coating materials used in these experiments were based on CoCrAlY alloys and also two NiCrAlTi compositions. The microstructural characteristics, freedom from "flake" defects etc. thickness and compositional control of the deposits were excellent. Results obtained from laboratory oxidation and high velocity corrosion rig tests on SIP-coated nickel alloy testpieces were encouraging.

Further work is required to explore the application of SIP to new coating compositions and also to investigate the possibility of increasing the deposition rate of high alloyed coating materials.

ACKNOWLEDGMENTS

The authors are indebted to P. Warrington, D. Rickerby and R.M.W. Hawes (AERE) for assistance with this work.

REFERENCES

1. Sykes, C. and Shirley, H.T. (1951) Special Report 43, ISI, London.

2. Hancock, P. (1961) Proc. 1st International Conference on Corrosion, Butterworth's, London. pp. 173-193.

3. Lewis, H and Smith R.A. Ibid pp. 213-232.

4. Dean, A.V. (1965) NGTE Report 267.

5. Hancock, P. (1968) Corrosion of Alloys at High Temperatures in Atmospheres consisting of Fuel Combustion Products and associated Impurities - a critical review, HMSO, London.

6. Bornstein, N.S. and Decrescente, M.A. (1971) TMS-AIME 2, pp. 2875-2900.

7. Stringer, J. (1972) Hot Corrosion in Gas Turbines, MCIC-72-08.

8. Hart, A.B. and Cutler, A.J.B. (eds.) (1973) Conference on Deposition and Corrosion in Gas Turbines, CEGB, Applied Science, London.

9. Condé, J.F.G. (1972) Specialist Conference on High Temperature Corrosion of Aerospace Alloys, Lyngby, Denmark, AGARD-CP-120 pp. 203-219.

10. Restall, J.E. (1975) NGTE Report 339.

11. Restall, J.E. (1980) Institute of Metallurgists Spring Residential Conference, Isle of Man, U.K. pp. 4/13-4/26.

12. NAVSEC (1974) Proc. 2nd US/UK Conference on Gas Turbine Materials in the Marine Environment, Castine, MCIC-75-27.

13. NAVSEC (1976) Proc. 3rd US/UK Conference on Gas Turbine Materials in the Marine Environment, Bath, U.K.

14. Coutsouradis, D. et al (eds.) (1978) Proc. COST 50 Conference Liege, High Temperature Alloys for Gas Turbines, Applied Science, London.

15. Restall, J.E. (1979) High Temperature Coatings for protecting Hot Components in Gas Turbine Engines. Metallurgia 46 pp. 676-685.

16. Restall, J.E. (1981) 'Surface Degradation and Protective Treatments', Ch. 10 of 'The Development of Gas Turbine Materials'. Ed. Meetham, G.W. Applied Science, London.

17. Restall, J.E. (1979) Improvements in or relating to nickel-chrome-base alloys. Brit. Patent Appl. 7925846.

18. Dugdale, R.A. (1977) Conference on 'Ion Plating and Allied Techniques'. Edinburgh, CEP Consultants, Edinburgh. pp. 177-186.

19. Coad, J.P. and Dugdale, R.A. (1979) Conference on Ion Plating and Allied Techniques, London, CEP Consultants, Edinburgh. pp. 186-196.

20. Mattox, D.M. and MacDonald, J.E. (1963) J. Appl. Phys. 34 pp. 2493-2494.

J F LANCHBERY and B WALTON

The application of protective oxide coatings to metals by rf sputtering and rf ion plating

SYNOPSIS

In this paper the basic physical processes involved in radio frequency (rf) sputtering and rf ion plating, in both inert and chemically reactive plasmas, are outlined. The advantages and limitations of the techniques in protective coating applications are discussed, with particular reference to the deposition of simple oxide, mixed oxide and glass coatings on metal surfaces.

THE AUTHORS

are with ERA Technology Ltd,
Leatherhead, Surrey.

INTRODUCTION

New and improved techniques for the vacuum deposition of thin solid films continue to be developed. Much of the stimulus for this work has been generated by the very rapid growth of the microelectronics industry where the controlled deposition and delineation of thin films of a wide variety of materials, often of very high purity, are fundamental to the fabrication of solid state electronic devices. It is now possible, in principle, to coat a flat solid surface with a thin solid film of controlled thickness and of almost any desired inorganic composition. The successful vacuum deposition of thin films, on a laboratory scale, onto the polished surfaces of sophisticated microelectronic devices, however, is quite a different matter from the widespread application of such techniques to the provision of substantial coatings on larger, perhaps irregular shaped, objects. Rapid deposition of films of low melting point metals, such as aluminium, can readily be accomplished by the relatively straightforward technique of vacuum evaporation from a heated source, and large scale coating by this method has been in use for many years. The process is, however, best suited to the coating of fairly flat surfaces and becomes more difficult to use if the melting point of the metal to be deposited is high or if films of complex alloys or chemical compounds are to be formed. Electron guns are now normally used to provide the intense heat necessary to evaporate refractory elements, but the risk of dissociation of chemical compounds, or preferential evaporation of alloy components on heating, remains a problem. For these reasons techniques which make

use of the sputtering process have, in spite of their inherent slowness, been adopted for the deposition of more refractory alloys and compounds. Sputtering techniques, in general, allow the deposition of thin films having the same chemical composition as the starting material.

The deposition of materials which are both refractory and electrically insulating presents a particular problem. Conventional sputtering techniques employing direct current (d.c.) applied fields cannot be used because the surface of the source material (the 'target') becomes electrically charged, effectively cancelling the applied field and extinguishing the plasma used for the sputtering process. Two methods of overcoming this problem are to use rf sputtering or ion beam sputtering. Ion beam sputtering is still being developed from a small scale laboratory process, whereas rf sputtering, which has been steadily developed over the last fifteen years, is suitable for large scale use. Powerful machines capable of depositing films over large areas have been built and rf sputtering is well established as a production method.

More recently the technique of ion plating, which combines some of the advantages of evaporation and sputtering, has attracted considerable attention, particularly for the deposition of metal films. Again, by employing an rf field for plasma generation the method may be used to deposit films of insulating materials. RF sputtering and rf ion plating are thus competing methods for the vacuum deposition of electrically insulating films. It is the principal purpose of this paper to discuss the main characteristics of the two techniques and, in particular, their suitability for the formation of metal oxide coatings on metal surfaces.

1 BASIC CHARACTERISTICS OF RF SPUTTERING AND RF ION PLATING

1.1 RF Sputtering

In all vacuum deposition techniques which make use of the sputtering phenomenon, material is removed from the source (usually called the 'target') by bombarding it with energetic gas ions.

The nature of physical sputtering is complex and not fully understood[1,2,3,4] but for the purposes of this discussion the process may be envisaged as one in which an energetic ion strikes a solid surface with sufficient momentum to remove one, or more, surface molecules. The minimum ionic energy needed to remove a molecule from a surface

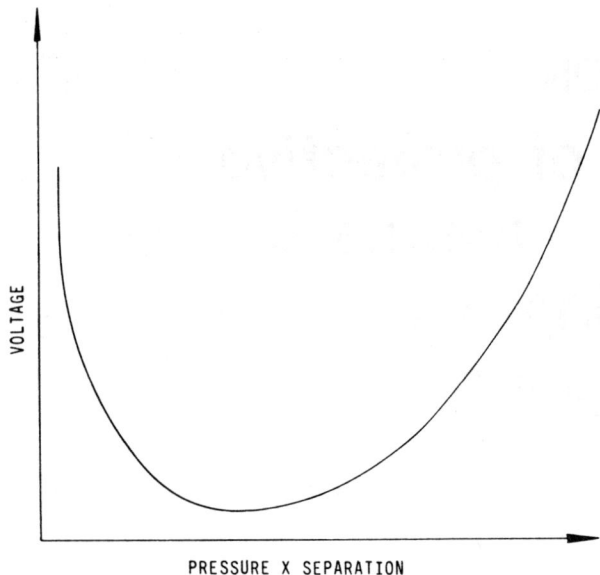

Fig. 1 Basic form of Paschen's 'Law' (voltage
required to form discharge is proportional
to gas pressure and plate separation)

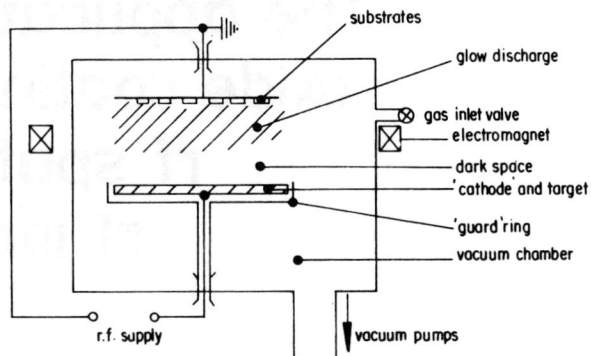

Fig. 3 Schematic: rf sputtering system

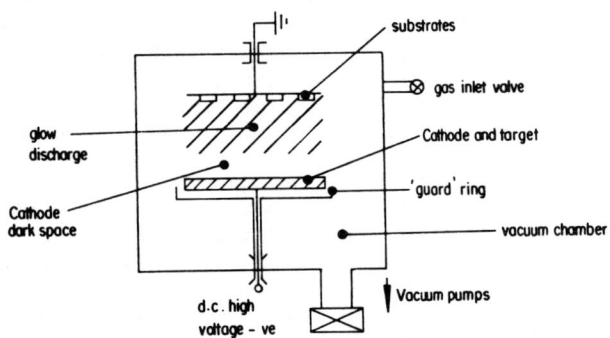

Fig. 2 Schematic: typical d.c. sputtering system

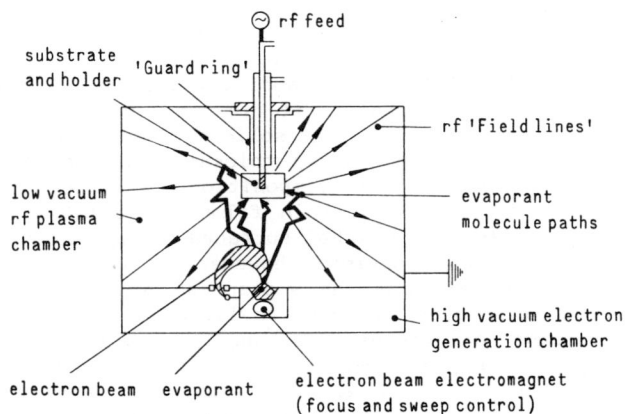

Fig. 4 Schematic: ion plating apparatus used
at ERA Technology Ltd.

(the sputtering threshold energy) is dependent both upon the solid material and the nature of the bombarding ion but is typically of the order of 10 eV[5]. The energy needed to achieve a sputtering yield of unity (i.e. one ionic impact yielding one sputtered molecule) may however be as high as several hundred electron volts. During sputtering the ejection of electrically neutral molecules is accompanied by the release of electrons, ions, and sometimes dissociated molecules. In most practical applications efforts are made to minimise such 'secondary' emissions.

The most convenient and widely used method of obtaining the energetic ions needed for sputtering is to remove ions from a gas discharge. Gas discharge plasmas are easily created by applying an emf between two parallel plate electrodes placed at least 2 cm apart in a vacuum chamber containing a gas at a fairly low partial pressure (typically 10^{-3} to 10^{-1} mbar). In the case of a d.c. applied field, if the gas pressure, emf and electrode separation are such that Paschen's 'Law' (represented schematically in Fig.1[6]) is obeyed then cold cathode electron emission occurs at the negative electrode and these electrons, accelerated by the electric field, cause partial ionisation of the gas and a plasma is 'struck' between the electrodes.

In order to obtain ions having energies significantly greater than sputtering threshold energies, the low power plasmas used in d.c. sputtering processes are operated in the 'abnormal cathode fall' mode where the whole cathode is surrounded by discharge and the voltage drop across the 'dark space' between the glow discharge plasma and the cathode is several hundred volts. The details of particle behaviour in this region are described, amongst others, by Hurley[7]. Similar dark spaces or 'cathode sheaths' can occur in low power plasmas where rf excitation is used[8] and high energy ions can thus be obtained in essentially the same manner.

In most vacuum deposition techniques employing plasma generated ions for sputtering of the film forming material, the material to be sputtered is placed at the cathode and the object to be coated (substrate) is placed at the anode. The sputtered molecules thus traverse the plasma prior to being deposited upon the substrate. A typical parallel plate sputtering system, utilising d.c. plasma excitation, is shown schematically in Fig.2, but other more complex electrode configurations are often used.

Except in the case of reactive processes, discussed later, the gases used for ionic generation in sputtering processes are the inert gases. These have the advantage of being most unlikely to react chemically with the sputtered material either at the electrode surfaces or in the plasma. Ideally, for maximum sputtering efficiency, the molecular mass of the bombarding gas ion should be similar to that of the material being sputtered[9] but in practice argon is very widely used because it gives a reasonable sputtering yield for most materials, and is cheap, readily available and very inert.

Although rf sputtering is superficially similar to d.c. sputtering, with an rf rather than a d.c. field being used, the physical mechanisms involved in the two techniques differ significantly. When a low frequency alternating current (a.c.) field is applied between two equal area parallel plates in a partial pressure of gas (as outlined above) then the resultant plasma simply fluctuates with the applied field to give a dark space at each electrode in turn. However, as the frequency of the field is increased it is observed[10] that the minimum pressure at which the plasma may be maintained decreases (i.e. Paschen's

'Law' is no longer of the same form). This effect begins to occur at frequencies above about 50 kHz and is due to the fact that the plasma, instead of being maintained by secondary electrons emitted from the cathode, is maintained primarily by electrons in the plasma which can ionise the gas. This occurs because, as the frequency increases the number of electron/gas collisions per unit time increases and, also, electrons suffering random collisions with gas molecules have a finite probability of gaining energy from the field, between successive collisions, as the field rapidly reverses. A proportion of electrons can thus gain sufficient energy from the field to ionise gas molecules and it is possible to maintain an rf plasma at a lower pressure than is possible with d.c. excitation. Also, if magnetic fields are used to confine electrons to the plasma, comparatively few electrons are needed to sustain the plasma and hence comparatively low field strengths may be used. In practice, of course, some electrons are always 'lost' and it is imperative to maintain a fairly high field in order to ensure that ions having sufficient energy to achieve high sputtering yields are generated.

It is found that in rf discharges between a symmetrical pair of metal plates the plasma is at a significant positive potential with respect to both electrodes, which each have a dark space. This occurs because plasma ions are considerably less mobile than electrons and many more electrons than ions therefore reach the electrodes during each half cycle at radio frequencies. This phenomenon, which was originally described in detail by Levitskii[11] and Tsui[12], evidently precludes the transference of one material preferentially from one plate to the other as both are sputtered. In practical rf sputtering systems one electrode is therefore capacitively coupled to the rf supply. A dark space then appears only at the capacitively coupled electrode, at which the sputtering target is placed, and the other electrode, which is often earthed, is maintained at a potential close to that of the plasma and receives the sputtered deposit.

Useful features of rf sputtering include the following:

(i) The technique allows the deposition of almost any solid inorganic compound. Although certain materials do dissociate to some extent when sputtered, this can be compensated for by introducing appropriate vaporised or gaseous species into the plasma, thereby reforming the original compound. This type of process is known as rf reactive sputtering.

(ii) RF sputtering systems can operate at far lower pressures than d.c. sputtering systems, typically in the range 10^{-4} to 10^{-2} mbar, where the mean free path of molecules, at 300k, is of the order of 1 cm. Gaseous inclusions in films are thus reduced.

(iii) Because rf plasmas which are contained by magnetic fields require comparatively little secondary electron excitation the number of secondary electrons incident at the substrate is reduced. Substrate heating due to electron bombardment is thus reduced and can be minimised further by confining electrons close to the cathode surface using magnets placed behind the cathode. This technique known as 'magnetron sputtering' produces enhanced ionisation of the plasma gas and hence increases the sputtering rate[13].

1.2 RF Ion Plating

Ion plating is essentially a process in which a vapour stream, usually from an evaporant source is deposited onto a receiving surface surrounded by a plasma.

In conventional d.c ion plating[14], the object to be coated is placed in a vacuum chamber, maintained at a pressure typically in the range 10^{-2} to 10^{-1} mbar, and forms part of a cathode assembly held at a high negative potential (commonly about 1 kV) relative to the rest of the chamber, which is usually 'earthed'. Cold cathode emission of electrons at the cathode is normally used to sustain the plasma, although hot electron emitters are sometimes used.

In a typical coating operation the substrate is first sputter etched to give an atomically clean surface. When the substrate has been sufficiently cleaned the coating material is introduced in vapour form from an evaporant source into the plasma. The vapour molecules undergo a large number of highly energetic collisions in the plasma, because the mean free paths of the ions and molecules is typically between 1 mm and 1 cm whilst the cathode to source distance is usually greater than 10 cm. (The majority of collisions in the chamber are of random momentum but in the abnormal cathode fall region, close to the substrate being coated, there is a net momentum transfer from the ions towards the cathode. In this region ionisation of evaporant is particularly likely to occur, hence the term 'ion plating', but the effect is usually quite minor and does not normally have an important effect upon film quality.) The evaporant is thus caused to impinge on the substrate over a very large solid angle and with a kinetic energy which may be as high as several hundred electron volts. The high energy of the coating molecules and the existence of an atomically clean substrate leads to the formation of highly adherent films. Moreover, very dense homogeneous films tend to form because sputtering continues throughout the period of film growth and causes the removal of atoms or molecules which are not tightly bound to the receiving surface.

RF ion plating is similar in operation to d.c. ion plating but the plasma processes are essentially those which occur in rf sputtering. (Fig.3 is a schematic diagram of an rf sputtering apparatus used at ERA Technology.) As with rf sputtering the high frequency field enables the deposition of electrically insulating compounds and the use of lower excitation voltages and gas pressures. RF ion plating therefore allows the deposition of a wider range of materials than d.c. ion plating[15] and also permits a greater range of control over both substrate etch and deposition parameters. RF reactive ion plating can be used not only to replace gaseous species 'lost' due to dissociation, as in rf sputtering, but also to form compounds during deposition. For example, aluminium may be evaporated into a plasma containing oxygen to form an aluminium oxide film.

2 RF SPUTTERING AND RF ION PLATING IN THE APPLICATION OF PROTECTIVE OXIDE COATING

RF excited plasma deposition techniques are particularly well suited to the deposition of metal oxide coatings. Indeed, rf sputtering and rf ion plating are probably the most versatile techniques currently available for the formation of oxide films because most oxides are either electrically insulating or poor semiconductors. A large number of oxides are extremely chemically inert, refractory, and both hard and mechanically durable. If applied as thin coatings they may thus protect metal surfaces against oxidation or reduction at elevated temperatures, prevent attack by corrosive chemicals, reduce erosion by fluid flow or wear by mechanical abrasion, and also provide electrical insulation. Some of the main advantages and limitations of rf sputtered and rf ion plated oxide coatings in protective applications are given below:

2.1 Range of Coating Materials

A considerably greater number of compounds may be deposited by rf sputtering than by rf ion plating because all current ion plating techniques use vacuum evaporation to vaporise the coating material. Thus many of the more complex oxides, particularly glasses, are often difficult to ion plate even with recourse to reactive plating. To take an extreme example, the authors have successfully formed 'doped' garnet films (of the type $Y_{3-x} Bi_x Fe_{5-y} Ga_y O_{12}$) by reactive sputtering. Since the basic unit cell comprises 160 atoms at highly specific sites, the deposition of such films by rf ion plating would, at the current stage of development of the technique, be quite impossible.

However, simple oxides are often suitable for metallurgical protective coatings. For example, some time ago the authors had a need to protect small areas (a few square centimetres) of AISI 316 stainless steel against the effects of exposure to high temperature in an oxidising atmosphere for periods in excess of 10^4 hours. There was an additional requirement that the protective films be electrically insulating (at least $10^{10} \, \Omega$ cm) at 600°. Both silica and a low alkali barium aluminosilicate glass of high electrical resistivity were successfully applied as coating materials by rf sputtering. Although the thermal expansion mismatch between the silica and the steel was greater than that between the glass and the steel the latter proved more suitable, principally because it was a more refractory material. The silica coatings protected and insulated the steel at temperatures in excess of the required level whereas the glass tended to soften and delaminate at temperatures above about 600°C. We have since found that it is possible to either rf sputter or rf ion plate silica onto stainless steel and that the sputtered films will withstand temperatures of at least 800°C. The rf ion plated films have yet to be fully evaluated but appear to have similar characteristics to sputtered films. It would be very difficult to prepare films of the barium aluminosilicate glass by ion plating.

2.2 Some Specific Properties of RF Sputtered and Ion Plated Films

Although both techniques allow film properties to be varied considerably there are certain characteristics which are peculiar to each process. For example, under optimum conditions the adhesion of ion plated films is invariably greater than rf sputtered films because of the 'self-cleaning' nature of the process. Film/substrate adhesion is, however, usually naturally high between most oxide films and metal surfaces so that the difference between rf sputtered and ion plated films may therefore be of little practical consequence in that case.

RF sputtered films deposited onto substrates at about 300k usually have either a very fine grained structure (with a grain size of perhaps a few hundred nanometres) or have no discernible structure. Microstructure can, however, be modified to yield larger grain size by deposition at elevated substrate temperatures or by application of electrical bias to the substrate. In most protective oxide coating applications fine grained or amorphous 'structures' are preferred as the films then tend to be less brittle. For example, the silicon dioxide films, mentioned previously, were deposited in vitreous form. Deposited as crystalline quartz the films would undoubtedly have cracked under the strains imposed by thermal mismatch.

In certain types of glassy coatings the applied films, although of the same nominal composition as the bulk material, may possess dissimilar properties. This is due to dissociation and reassociation during deposition and presumably to some rearrangement on an atomic scale. For example, we have deposited films of Corning 7059 glass which in bulk form is brittle and fractures at low strain. When rf sputtered to form thin films (analysed to be of the same chemical composition as the target) the glass withstood applied strains of up to 1.5% at room temperature.

In general, rf sputtering is ideally suited to the deposition of glassy materials and such coatings have been used, at ERA Technology, for a wide variety of experimental applications from the provision of integral covers upon silicon solar cells to the protection of metals for use in the cores of nuclear reactors.

The degree of control which is available over the important deposition parameters during rf ion plating enables films of almost any microstructure to be obtained. Because loosely bound material is removed from non-preferred lattice sites it is possible to obtain highly crystalline films. However, because of limitations in the range of compounds which may be deposited this has only been achieved for simple compounds, when reactive sputtering may sometimes be used to advantage. Also, because fine grained or amorphous oxide films are generally preferred for protective metallurgical coatings, emphasis has been placed upon the formation of such coatings. We have found that SiO_2 and Al_2O_3 films exhibit good high temperature performance when deposited in nearly amorphous form and also that such films will withstand exposure to highly corrosive liquids. Hot ferric chloride, concentrated hydrochloric and nitric acids have so far been shown to have no effect upon metals protected in such a manner.

2.3 Types of Substrate Which May Be Coated

RF ion plating has considerable advantages over rf sputtering in that it may be used to coat highly irregular shaped objects and substrates which are prone to thermal degradation.

The 'throwing power' (i.e. ability to uniformly coat the entire surface of irregular shaped objects) of ion plating is the mean free path (mfp) of the molecules in the plasma. During rf ion plating work performed by the authors, at a pressure of 10^{-3} mbar which corresponds to a mfp of several centimetres it has proved possible to ion plate into recesses having maximum apertures of less than 1 mm. We are currently performing work which should enable further quantification of the relationship between throwing power and pressure to be made.

On the other hand the conventional parallel plate configuration of an rf sputtering machine is well suited to the deposition of coatings of uniform thickness flat surfaces. It can be arranged that the area of the receiving surface corresponds almost exactly with that of the sputtered target. Under these circumstances a high proportion of the sputtered material is utilised in forming the coating and it can be readily applied to the substrate material in the form of continuous sheet.

Thermal degradation of substrates is not normally a major problem in most metal coating work

but it may be critical in some applications, for example, where the substrate is very thin. In rf sputtering thermal effects at the substrate are caused primarily by bombardment by secondary electronics emitted from the cathode. This does not arise in ion plating and although high ionic bombardment rates can give rise to surface heating this can normally be kept to a low level if necessary.

We have recently performed work in which the relative substrate heating effects of the two processes were directly compared. In these experiments silica films were deposited, by both techniques, onto sheets of organic material which was known to degrade severely at temperatures in excess of $80°C$. The rf sputtering apparatus was operated at a significantly lower plasma power than the rf ion plating apparatus and yet the substrates coated by sputtering exhibited severe degradation whereas substrates which had been ion plated were unaffected.

2.4 Scale and Cost

Both rf sputtering and rf ion plating can be scaled up for volume production. For coating a large number of separate components rapidly, rf ion plating is probably the preferred method as the deposition process can be very fast. The authors have found that it is possible to form refractory oxide films, in an experimental ion plater, at growth rates of greater than 1 m min^{-1}, whereas even in large sputtering machines used for production purposes the rates achieved for the deposition of oxide films are normally in the range 1 to 10 m hr^{-1}.

It has been reported that d.c. ion plating has been used in preference to electroplating for batch coating aircraft components with protective metal layers[16]. Small items have been coated at rates of about 25,000 hr^{-1} and larger items, such as aircraft landing gear and jet engine stator vane assemblies, have also been coated. The cost of operating such systems, which utilise vacuum chambers of 6 feet in diameter by 10 feet long, was reported to be about $3.30 per hour (exclusive of labour) at the end of 1980. Less than one operator was required to run such a machine. The cost of running large rf ion plating systems would be expected to be similar.

Although rf sputtering is a lower rate deposition process it is ideally suited to continuous coating of flat sheet and production machines have been built for this purpose. Sputtering machines can have target material utilisation factors as high as 90% and continuous coaters require minimal supervision. The cost per hour of using such machines is therefore lower than for ion platers. High rate batch sputtering processes for coating automobile parts with metals have been developed and these are extensively used in the Japanese motor industry. Obviously operating costs must be low enough to justify the use of such techniques in such a competitive environment.

In general, although the initial capital cost of all vacuum deposition equipment is relatively high, running and materials costs are low. Thus rf sputtering and rf ion plating are likely, in the future, to be used to provide protective oxide coatings for metals on a production scale.

CONCLUSIONS

RF sputtering and rf ion plating are useful methods for the deposition of thin protective oxide coatings on metals. Sputtering is more suitable for the deposition of complex compounds or mixed oxides of controlled composition and can be conveniently used to coat flat surfaces. Ion plating is faster and better suited to coating irregular shaped components.

REFERENCES

1. Anderson, H H: Appl. Phys. Phys. 18 (1979) p.131.

2. Etzkorn, H W; Littmark, U and Kirschner, J: in P Varga, Betz, G and Viehbock, F P (Eds.) Symp. on Sputtering, Vienna (1980), Institut für Allgemeine Physik, Vienna, 1980, p.542.

3. Tsaur, B Y; Matteson, S; Chapman, G; Liau, Z L and Nicolet, M.-A: Appl. Phys. Lett. 35 (1979), p.825.

4. Carter, G and Armour, D G: Thin Solid Films, 80, No.1/2/3 (1981) pp.13-29.

5. Wehener, G K and Anderson, G S, in Maissel, L I and Glang, R (Eds.): Handbook of Thin Film Technology, McGraw-Hill (1970) NY, Chap.3 pp.15-20.

6. Vossen, J L and Kern W (Eds.): Thin Film Processes, Academic Press (1978) NY.

7. Hurley, R E: Thin Solid Films, 86, No.2/3 (1981) pp.241-253.

8. Davidse, P D and Maissel, L I: Trans. 3rd Intern. Vacuum Congr., Stuttgart (1965).

9. Wehner, G K and Anderson, G S: Op. cit. Chap.3, pp.2 and 3.

10. Acton, J R and Swift, J D: Cold Cathode Discharge Tubes, Academic Press (1966) NY.

11. Levitskii, S N: Sov. Phys. Tech. Phys. English Trans. 27 (1957), p.913.

12. Tsui, R T C: Phys. Rev. 168 (1968).

13. Nyaiesh, A R: Thin Solid Films, 86, No.2/3 (1981), pp.267-277.

14. Mattox, D M: Electrochem. Technology, 2 (1964) pp.295-298.

15. Davey, J G and Hanah, J J: J. Vac. Sci. Tech. 11, No.1, (1974) pp.43-46.

16. Steube, K E and McCrary, L E: J Vac. Sci. Tech. 11, No.1, (1974) pp.362-365.

K H KLOOS, E BROSZEIT, H M GABRIEL, and H J SCHRÖDER

Thin ceramic coatings deposited onto nodular cast iron by ion and plasma assisted coating techniques

SYNOPSIS

Sputtering and ion plating, two potential techniques to produce hard and wear resistant coatings, were used to deposit thin TiN coatings onto nodular cast iron.
Measurements of the tribological properties related to the coating parameters are reported and compared to uncoated and electrochemical Cr-plated samples.

THE AUTHORS

are with TH Darmstadt,
Inst. f. Werkstoffkunde,
6100 Darmstadt, Germany

1. Introduction

TiN as material for wear protection applications has been known for several decades, but the possibility of depositing thin TiN coatings on tools and structural components has only existed for about 20 years. The development of the chemical vapour deposition process in the sixties and the first deposition of TiC onto cemented carbide tool tips started the commercial application of hard and wear resistant coatings like TiN, Ti(C, N) and also multilayers (1-5).
The high temperatures which are generally required in CVD processes do not affect high temperature materials like cemented carbides and Stellites® . However, tool steels show an enormous decrease in hardness when exposed to temperatures of over 550°C and therefore heat treatment has to be delayed. This is one of the major disadvantages of the chemical vapour deposition processes. Therefore a growing interest can be seen for the use of physical vapour deposition processes like sputtering and ion plating. These two deposition processes allow the deposition of nearly any material onto materials which cannot resist high temperatures because of their structure and/or their treatment.

During recent years different papers have dealt with the sputtering and ion plating of TiN onto tool steels particularly onto drills and cutters (6-18).
These coatings were deposited in the so-called "reactive mode" where titanium is sputtered or ion plated in a N_2-atmosphere.
In most publications, however, TiN was deposited onto very high quality materials (highly alloyed steels, sintered cemented carbides).
For economic reasons also, cheap materials should be coated with thin, functional coatings, where the bulk material provides the strength and rigidity and the coating serves as a wear and/or corrosion protection.
The subject of this investigation is the deposition of hard, wear resistant coatings onto cast iron by sputtering and ion plating.

2. Experimental details

The sputtered coatings were deposited by using a turbomolecular-pumped sputtering unit (Leybold Heraeus Z 400) with a magnetron source and an rf-supply. The substrates were sputter cleaned prior to deposition. The cleaning and coating parameters are listed in Table 1.
The coatings were deposited at different N_2-partial pressures and bias voltages.
The ion plated coatings were deposited in a Leybold A 700 Q ion plating unit. This unit is specially modified for ion plating processes i.e. the 10 kW e-beam gun (Pierce-type, 90° beam-deflection) used for Ti evaporation is differentially pumped.
The plating parameters are listed in Table 2.
The thickness of the sputtered and ion plated coatings is 1 μm. The investigation included uncoated and Cr-plated specimens. The thickness of the electroplated Cr-coatings is about 200 μm and this type of composite material is used for industrial applications at the present time.
The wear tests were performed with a modified pin-on-disk machine under pure sliding conditions at mixed friction.

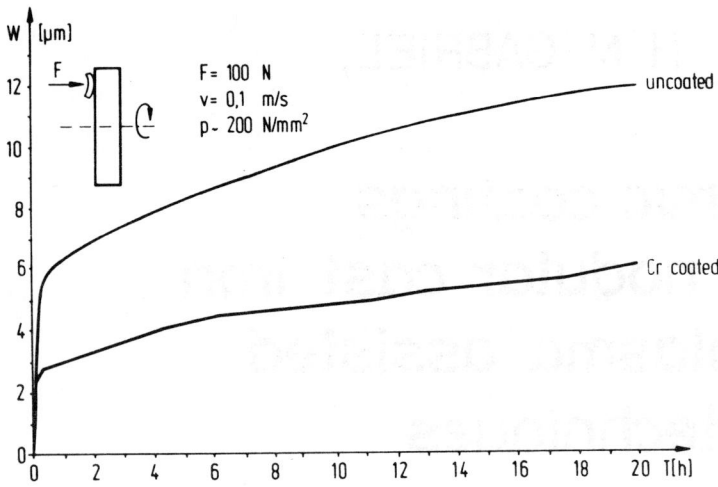

1   Wear behaviour of uncoated and Cr
    plated cast irons

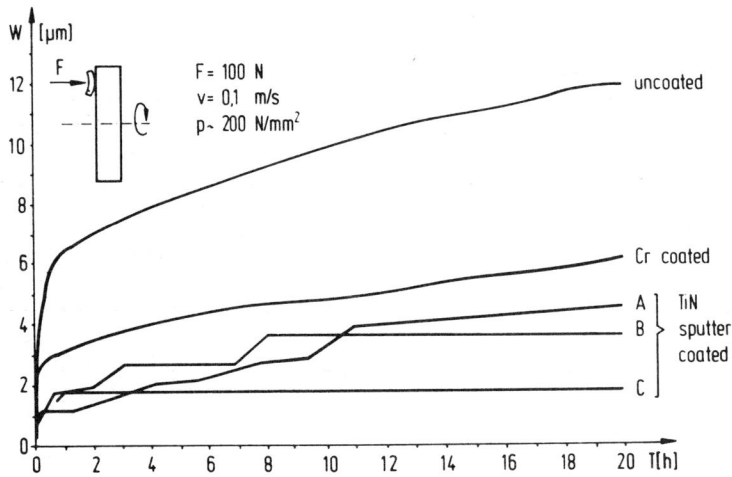

2   Wear behaviour of TiN sputter coated
    cast iron

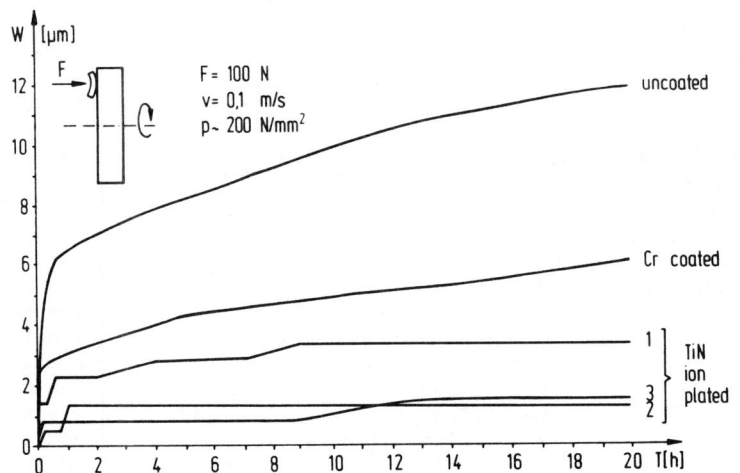

3   Wear behaviour of TiN ion plated
    cast iron

Table 1 PARAMETERS FOR SPUTTERED COATINGS

Sputter cleaning parameters

$t = 2$ (min) $U_{DC} = 800$ (V)

$p = 2 \cdot 10^{-2}$ (mbar) $I = 100$ (mA)

Deposition parameters

charge nr.	A	B	C	D	E	F
deposition rate(Å/s)	25					
Voltage (d.c.) (V)	380			210		
current (mA)	360			380		
N_2 partial pressure (10^{-4} mbar)	1,5	1,8	2,0	1,8		
bias voltage (V)	0			20	40	60

Table 2 PARAMETERS FOR ION PLATED COATINGS

Sputter cleaning parameters

$t = 5$ (min) $U = 5$ (kV)

$p = 4 \cdot 10^{-2}$ (mbar) $I = 330$ (mA)

Deposition parameters

charge nr.	1	2	3
deposition rate (Å/s)	20	20	50
substrate potential (kV)	5	4	4
chamber pressure (10^{-2} mbar)	6	6	6,5

An engine oil (Shell Rotella 20 W 20) was used as a lubricant at a temperature of 50°C.
The wear tests were performed at the following conditions:

load	100	N
sliding speed	0.1	m/s
initial contact pressure	200	N/mm^2
time	20	h

The material of the counterspecimen (disk) and the specimen to be coated was cast iron of the following composition:

3.2 - 3.5 % C,   1.8 - 2.2 % Si,   0.6 - 1 % Mn
0.3 - 0.5 % P,   max. 0.07 % S ,   0.2 - 0.5% Cr.

3. Results

3.1. Wear and friction behaviour of Cr-plated and uncoated specimens

The wear behaviour of the uncoated and Cr-plated specimens is shown in fig. 1.
After a short running-in-period with a very steep wear-time gradient the uncoated specimen gave a continuous wear increase with a smaller slope over the whole test duration.
A similar behaviour was observed with the 200 μm Cr coating but with considerably less wear in total.
The coefficient of friction of the uncoated specimen is slightly lower than the Cr-plated specimen but it shows more deflections.
The friction coefficients are listed in table 3a.
The total wear in each test (shown in Fig. 1-3) is the sum of the wear of the coated or uncoated specimen and the wear of the counterspecimen. (disk). The wear of the disk was detected by profile measurements. Wear for the tests with the uncoated and Cr-plated specimens varied between 0.6 and 0.9 μm (table 3b).

3.2. Wear and friction behaviour of sputtered TiN coatings

Fig. 2 shows the wear behaviour of sputtered TiN-coatings. These coatings were deposited at different N_2 partial pressures while the other parameters remained constant.
A decrease of wear was noticed with the increase of the N_2 partial pressure. The reason might be an increase of hardness, but the golden yellow colour of TiN coatings - a quite good indicator for the stoichiometry and the hardness - did not change in all the different charges. There was no possibility of measuring the hardness of such thin coatings.
The TiN coatings show a different wear behaviour compared with the uncoated or Cr-plated specimens.
After a short running-in period a steady state is reached followed by a repeated wear increase and a transition in the steady state (charges A and B).
Charge C, however, shows no further wear increase after the running-in period and then a steady state is reached.
The coefficients of friction (Tab. 3a) of the sputtered coatings are higher at the early stage of the wear test and decrease with the test duration. At the end of the test, charges B and C show the same μ as the uncoated specimen.
The wear of the counterspecimen is 1.5 μm for charge A, 1.2 μm for charge B and 1.4 μm for charge C (table 3b). This shows that the application of TiN protects the coated specimen but the uncoated counterspecimen is worn twice as much as the uncoated friction couple.
Different bias voltages were applied to the substrate to improve the properties of the sputtered coatings.
However, a significant wear increase was noted and SEM micrographs revealed a very flaky structure.

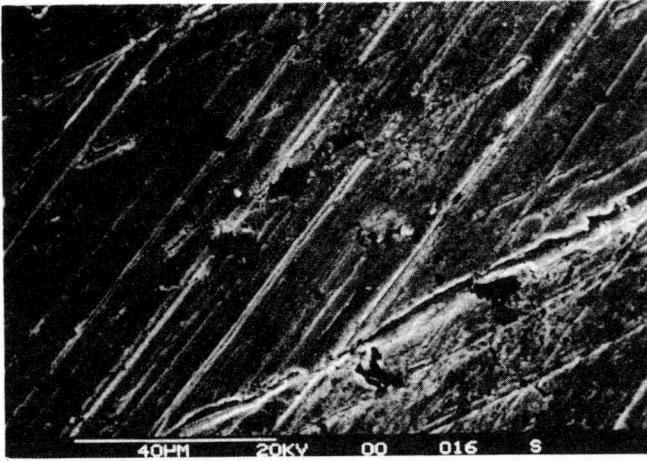

4 SEM micrograph of an uncoated cast
   iron specimen

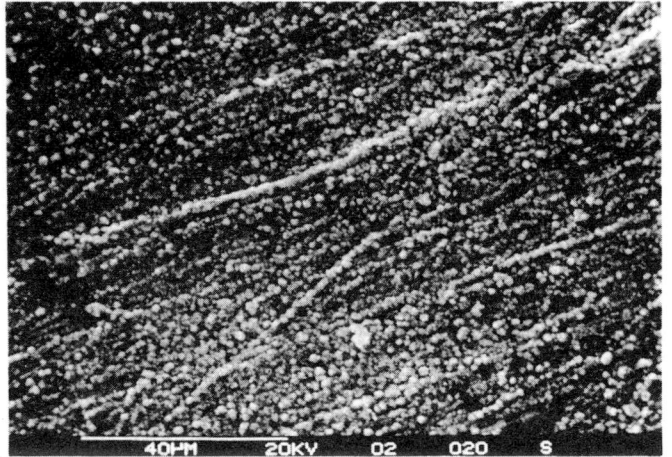

6 Ionplated TiN coating

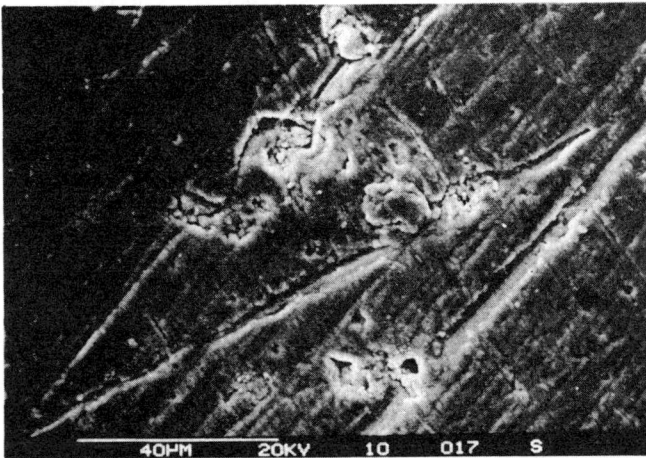

5 Sputtered TiN coating

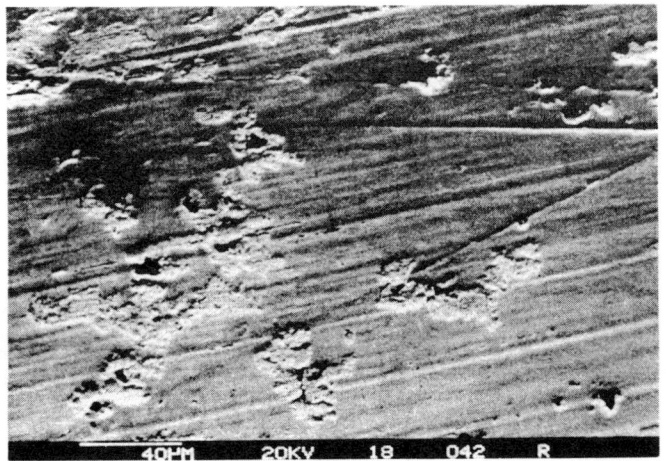

7 Worn TiN sputter coated surface

Table 3a FRICTION COEFFICIENTS OF UNCOATED,
Cr- AND TiN-COATED CAST IRONS

Table 3b WEAR (μm) OF THE UNCOATED
COUNTERSPECIMEN AFTER 20 h

Time (h)	start	2	5	10	20	
uncoated	0.09	0.10	0.10	0.08	0.08	0.6 - 0.9
Cr plated	0.11	0.12	0.11	0.09	0.09	0.8
TiN sputtered A B C	0.14 0.13 0.15	0.16 0.13 0.12	0.14 0.10 0.10	0.13 0.06 0.09	0.10 0.08 0.08	1.5 1.2 1.4
TiN ionplated 1 2 3	0.10 0.11 0.09	0.10 0.10 0.09	0.10 0.10 0.08	0.07 0.10 0.10	0.08 0.10 0.08	1.2 1.2 - 1.4 1.3 - 1.5

3.3 Wear and friction behaviour of ion plated
TiN coatings

Fig. 3 shows the wear behaviour of ion plated
TiN coatings. The wear of charge 1 specimen
is characterized by a repeated change of wear
increase and steady state condition.
Charge 2, however, shows a steady state after
a short running-in period. There is no additional wear increase even at rather high combined
compressive and shearing stresses.
Similar results were obtained for charge 3. All
3 charges show a nearly constant coefficient of
friction of about 0.1 (Tab. 3a).
The counterspecimen's wear is about 1.3 μm.
That is twice as much as that of the uncoated
friction couple (Tab. 3b).

4. Discussion

Fig. 4 shows a SEM micrograph of a finally
treated but uncoated cast iron sample. The cast
iron has a rather heterogeneous structure and
the nodular graphite penetrates the surface
(black areas). Also there is a strong orientation of grooves from the final grinding operation.
Figs. 5 and 6 show the surfaces of sputtered and
ion plated TiN coatings.
The coatings do not smooth the surface. The
sputtered coatings show rather big defects but
a very dense and smooth structure, while the ion
plated coating shows a very grained structure.
SEM investigations revealed that graphite nodules
are only partially coated and these areas are
possible defects for later detachment of the
coatings.
The coatings sputtered at different N_2 partial
pressures are smoothed by the frictional
stress and partially detached due to the
defects caused by the penetrating nodules
where the adhesion of the coating is very poor
(Fig. 7).
Fig. 8 demonstrates the peeling of a sputtered
coating by forming bubbles.
Fig. 9 shows a worn ion plated TiN coating
(charge 1). The failure of the coating is also
caused by the heterogeneous structure of the
cast iron but by improving the plating parameters (charge 3) detachment is minimized
(Fig. 10) and the coatings are able to withstand the combined compressive and shearing
stress (Fig. 10).
Possible defects are stopped by grinding
grooves and the neighbouring coating is fully
intact (Fig. 11).
As is shown by the SEM micrographs
(Fig. 5, 6) the sputtered and ion plated
coatings differ widely in structure. The ion
plated TiN coatings feature a very nodular
structure compared to the sputtered coatings
and generally a nodular grained coating has a
poor cohesion. However, as long as the coatings
are very thin the performance of the coatings
is more greatly influenced by adhesion than
cohesion. The coatings in general give good wear
resistance to the protected specimen, but they
have a deleterious influence on the wear of the uncoated counterspecimen. In all the tests with
the TiN coated specimens the wear of the disks
is about twice that of the uncoated friction
couple.
This is due to the rather low contact pressure
which remains when the uncoated samples are run
in.
For the tests with the coated samples the contact pressure is higher due to the decreased wear
and the hard TiN coatings partially worn on defect
areas give high abrasion because of their sharp
edges as demonstrated in SEM micrographs
(**Figs.** 12, 13).

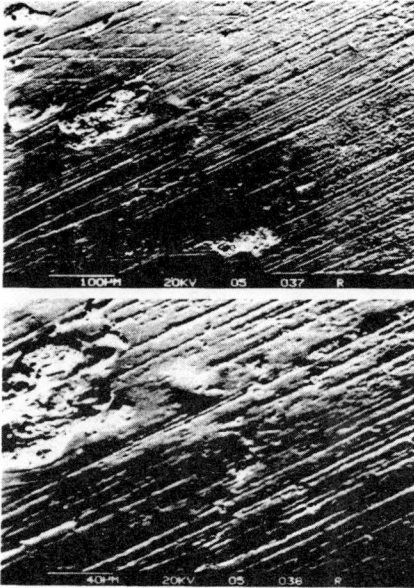

8 Worn TiN sputtered coating

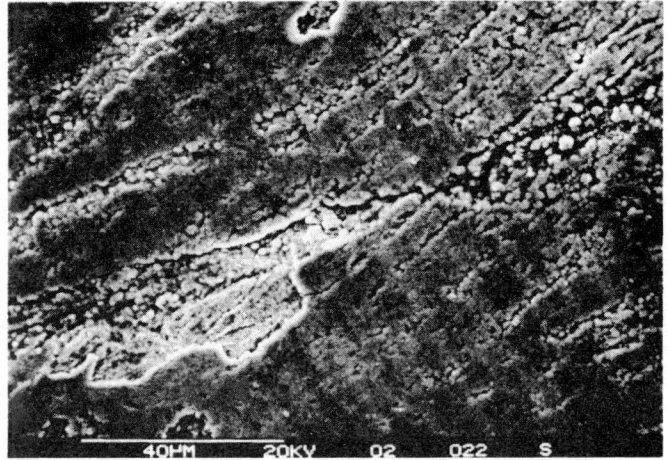

10 Worn ion plated TiN coating (charge 3)

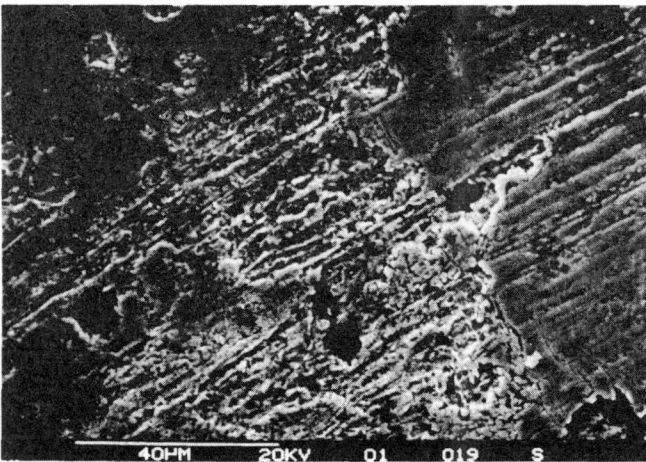

9 Worn ion plated TiN coating (charge 1)

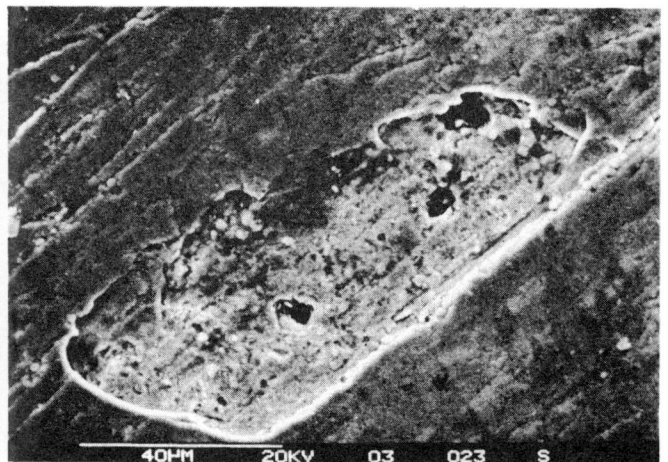

11 Coating detachment on areas with high concentrations of graphite nodules

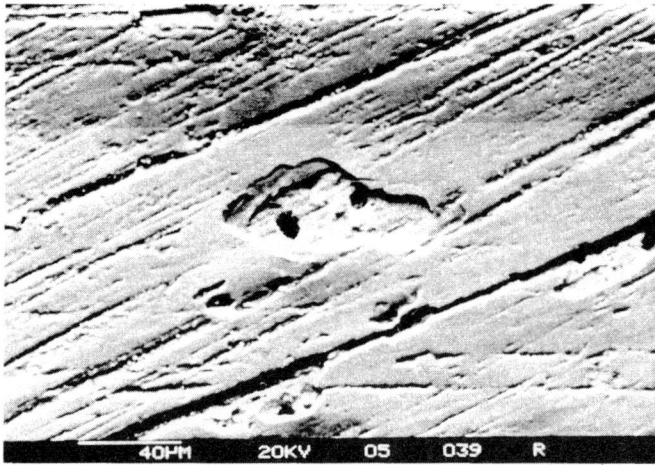

12 Worn TiN coating

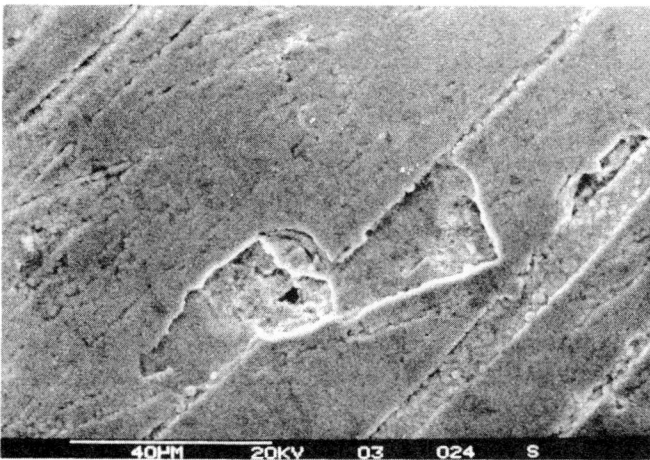

13 Worn TiN coating

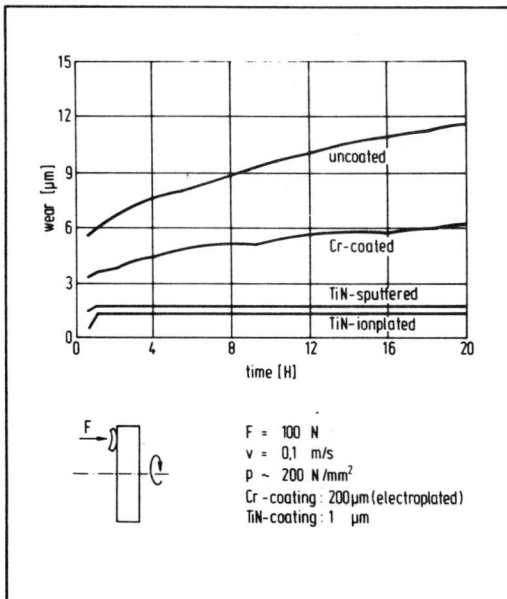

14 Wear behaviour of coated and un-
    coated cast irons

5. Summary

The deposition of hard and wear resistant TiN coatings onto cast iron - a cheap and easily available material - by sputtering and ion plating shows a considerable wear decrease in tribological tests (Fig. 14).
The heterogeneous structure of cast iron is responsible for the high defect concentration in the coatings. This is the reason for partial coating detachment but without any failure of the tribocontacts.
By an optimization of the sputtering and ion plating parameters the wear resistance of the coatings and their adhesion to the very heterogeneous bulk material can be greatly influenced.

ACKNOWLEDGMENT

The authors wish to thank the Deutsche Forschungsgemeinschaft (DFG) for their financial support.
The paper is a contribution to the "Sonderforschungsbereich 152 Oberflächentechnik" of the DFG.
The work is part of the Dr.-In.- dissert. of H. M. Gabriel.

REFERENCES

1. Ruppert W.
   Die Abscheidung von Titankarbidüberzügen auf Stahloberflächen
   Metalloberfläche 14. Jg. 1960 Heft 7

2. Ruppel W., Schlamp G.
   Werkzeuge mit Titankarbid-Überzug
   Bänder, Bleche, Rohre Okt. 1964

3. Takahashi T., Sugiyamak, Tomita K.
   The chemical vapor deposition of TiC coatings on iron
   Electrochemical Science, Dec. 1957, 1230 - 1235

4. Schintlmeister W., Pacher O.
   Titankarbid und Titannitrid für hochverschleißfeste und dekorative Schichten
   Metall 28 (1974) 690 - 695

5. Pierson H. O.
   Titanium carbonitrides obtained by chemical vapor deposition
   Thin Solid Films, 40 (1977) 41 - 47

6. Fleischer W., Schulze D., Wilberg R., Lunk A., Schrade F.
   Reactive Ion Plating with auxiliary discharge and the influence of the deposition conditions on the formation and properties of TiN Films
   Thin Solid Films, 63 (1979) 347 - 356

7. Matthews A., Teer D. G.
   Ion plated TiN coatings for dies and moulds
   Proc. IPAT 79 London July 1979

8. Zega B.
   Reactive deposition - a challenge for ion plating and allied techniques
   Proc. IPAT 79 London July 1979

9. Aronson A. J., Chen D., Class W. H.
   Preparation of titanium nitride by a
   pulsed dc magnetron reactive deposition
   technique using the moving mode of
   deposition
   Thin Solid Films, 72 (1980), 535 - 540

10. Ramalingam S., Winer W. O.
    Reactively sputtered TiN coatings for
    tribological applications
    Thin Solid Films, 73 (1980) 267 - 274

11. Chevallier J., Chabert J. P.
    Microhardness of TiN coatings obtained
    by reactive cathodic sputtering
    Thin Solid Films, 80 (1981) 63

12. Buhl, R., Pulker H. K., Moll. E
    TiN coatings on steel
    Thin Solid Films, 80 (1981) 265 - 270

13. Münz W., Heßberger G.
    Herstellung von harten Titannitrid-
    schichten mittels Hochleistungszerstäuben
    Werkstoffe und ihre Veredelung, Heft 3,
    1981

14. Gühring K., Kerschl W.
    Hartstoffbeschichtete Schneidwerkzeuge
    aus Schnellarbeitsstahl
    Industrieanzeiger Nr. 100, 12/1980

15. Zega B., Kornmann M., Amiguet J.
    Hard decorative TiN coatings by ion
    plating
    Thin Solid Films, 45 (1977) 577 - 582

16. Nimmagadda R. R., Doer H. J., Bunshah R. F.
    Improvement in tool life of coated high
    speed steel drills using the activated
    reactive evaporation process
    Thin Solid Films, 84 (1981) 303 - 306

17. Matthews A., Teer D.G.
    Deposition of Ti-N compounds by thermi-
    cally assisted triode reactive ion plating
    Thin Solid Films, 72 (1980) 541 - 549

18. E. Sirvio, M. Sulonen, H. Sundquist
    Abrasive wear of ion plated TiN coatings
    on plasma nitrides steel surfaces
    Intern. Conf. on Metallurgical coatings
    and process technology
    April 5 - 8, 1982; San Diego, California

R GILLET, A AUBERT, and A GAUCHER

Hard chrome coatings deposited by physical vapour deposition

SYNOPSIS

Coatings of pure chromium and of carbon or nitrogen doped chromium have been prepared by cathodic magnetron sputtering of a Cr target in atmospheres of argon, argon and methane, or argon and nitrogen.

The Cr,C and Cr,N coatings have Vickers hardnesses up to 3500 kgf/mm^2 and possess fretting-wear, fatigue and corrosion properties superior to those of electrolytic hard chrome. X-ray diffraction analysis revealed the presence of supersaturated solid solutions of C and N in Cr for concentrations of C and N up to several weight %, with no detectable nitride or carbide formation as would be predicted by binary Cr-C and Cr-N phase diagrams

THE AUTHORS

R.Gillet, A.Aubert, CENG/DMG/LEMM, B.P. 85 X, 38041 Grenoble cedex, France
A.Gaucher, HEF, 42160 Andrézieux-Bouthéon, France

1. INTRODUCTION

Thick coatings of electrolytic hard chrome are frequently used in industrial applications but the following shortcomings limit their use :
- hydrogen embrittlement in steels, particularly high resistance steels, caused by the electrolytic process,
- low deposition rates,
- poor throwing power,
- poor electrical efficiency of deposition (about 15 %),
- pollution problems associated with the Cr plating bath.

To avoid such problems, physical vapour deposition such as vacuum evaporation or cathodic sputtering can be considered for the deposition of similar coatings.

This paper reports the results obtained for pure and doped chromium coatings deposited by high speed magnetron sputtering.

2. EXPERIMENTAL PROCEDURE

2.1 Equipment

The depositions were carried out in a cubic evacuated chamber (1 m^3) equipped with a 100 mm x 200 mm planar magnetron cathode with the substrate and cathode in a vertical configuration. Up to 5 cylindrical substrates could be coated at the same time using a device which simultaneously rotated all 5 substrates in a plane parallel to the target.

The samples were heated by a resistance heater located in front of the target, and their temperatures were monitored by a chromel-alumel thermocouple.

2.2 Chromium Target

The Cr target was forged under vacuum and contained the following levels of impurities:

$O_2 \leqslant 200$ ppm $C \leqslant 1400$ ppm $N \leqslant 100$ ppm
$Fe \leqslant 14\ 700$ ppm $S \leqslant 1400$ ppm

2.3 Gases

The gases used were as follows :
- Argon - spectrographic grade $O_2 \leqslant 0.7$ vpm
                              $H_2 \leqslant 1$ vpm
                              $CH_4 \ncong 0$
- Reactive gases - N45 grade O_2, N_2 and CH_4.
The residual gas pressure prior to argon injection was 5×10^{-7} torr.

2.4 Test pieces

The sample substrates were either planar plates or cylindrical pieces of such steels as 304 L, 038 and 4135, and glass plates were used as reference substrates. In general, the substrates were chemically degreased and pickled and in some cases cleaning by ionic bombardment was also utilised.

2.5 Operating Conditions

In all cases, the chromium deposits were sputtered in a constant partial pressure of argon of 2×10^{-3} torr and with a constant cathodic discharge power of 2 kW. The target to substrate spacing was 50 mm, and the substrate temperatures were 200° C, 300° C or 400° C. The substrates were negatively biased to -100 volts. The deposition rates varied between 0.4 and 0.65 μm/minute depending on the reactive gas pressure.

3. RESULTS AND DISCUSSIONS

3.1 Thickness

Coating thickness was determined for very thin films on glass substrates with a special probe

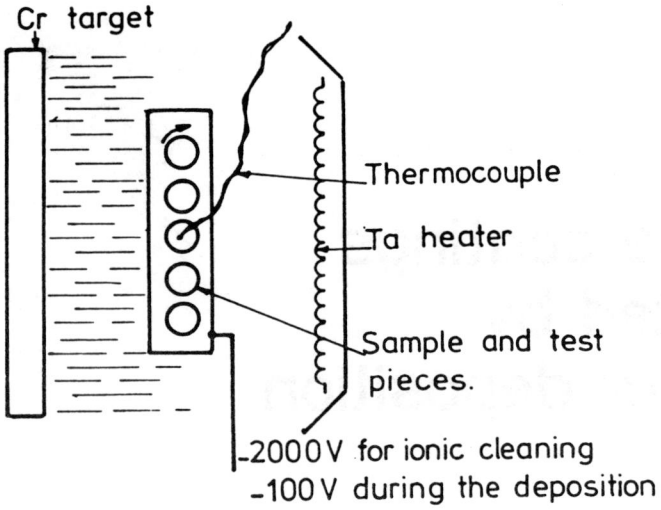

1 – Target – substrate configuration

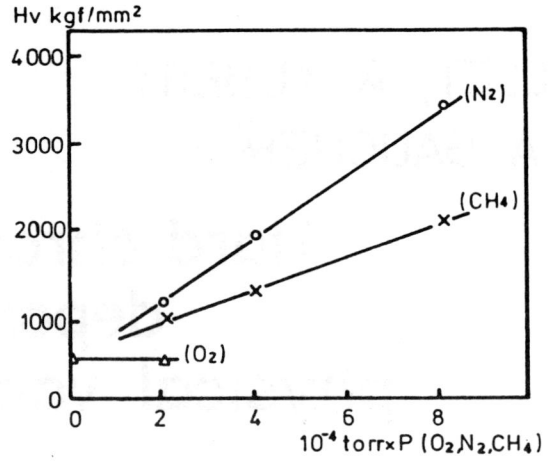

2 – Deposit thickness distribution on a fatigue test piece

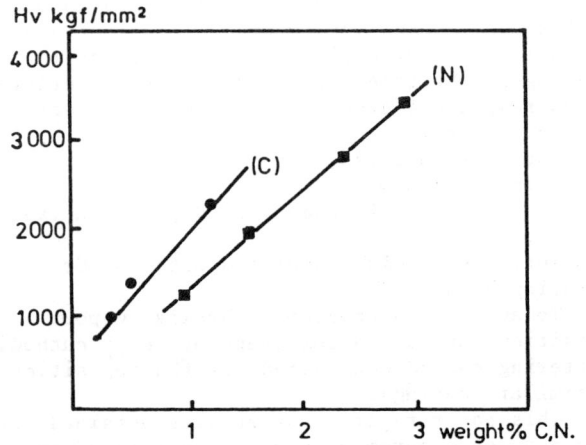

3 – Hardness as a function of partial pressure of reactive gas

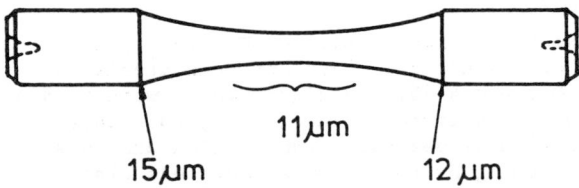

4 – Hardness as a function of C and N concentration in the Cr matrix

type TALYSURF instrument and a metallographic microscope was used for thicker deposits. Figure 2 shows the distribution of deposit thickness on a fatigue test piece.

3.2 Hardness

For pure chromium deposits, the microhardness was observed to change as a function of substrate temperature and varied between 425 and 840 kgf/mm^2. For Cr,O deposits the microhardness was independent of the O$_2$ content in the chromium matrix. For Cr,C and Cr,N, figures 3 and 4 show respectively the variations in microhardness as a function of the partial pressure of the reactive gas and as a function of the C and N concentrations in the chromium matrix.

3.3 Morphology

The morphologies of coating fractures were observed by optical and scanning electron microscopes. For pure chromium deposits these examinations revealed a columnar structure perpendicular to the substrate surface with a tendency to nodule growth initiated at surface defects. Figure 5 shows an optical micrograph of a pure chromium coating on a stainless steel substrate and figure 6 shows a fractograph of the same coating. The grain size varied as a function of substrate temperature and also depended on the surface curvature (planar or cylindrical).

The Cr,N coating was very dense and without columnar structure as shown in figure 7. Figures 8 and 9 show the variations in the morphology of the Cr,C coatings as a function of carbon concentration in the deposit. These coatings were very dense with small grain sizes.

3.4 Structure and Composition

A number of authors (ref. 1) have reported the formation of carbides and nitrides of chromium during the reactive sputter deposition of chromium in atmospheres of argon-acetylene and argon-nitrogen.

However, in the present investigation, X-ray diffraction analysis did not detect carbides or nitrides but only supersaturated solid solutions of C and N in the chromium matrix.

The quantitative analysis of C and N by glow discharge indicated a 1.2 weight percent concentration of carbon, and 2.9 weight percent for nitrogen in Cr,C and Cr,N deposits respectively. These values differ from those predicted by the Cr-C and Cr-N binary phase diagrams which give much lower limits for the solubilities of C and N in chromium.

Thus it appears that with good operating conditions it is possible to obtain Cr,C and Cr,N coatings with compositions and structures which are theoretically out of thermodynamic equilibrium. Such structures are thermally stable up to 500° C but at higher temperatures carbide and nitride formation probably takes place with consequent variations in coating hardness.

X-ray diffraction also showed the deposits to have highly oriented textures.

3.5 Corrosion Resistance

Coatings of pure Cr ; Cr,O ; Cr,C and Cr,N were exposed to the standard NFX 41002 salt spray test. For these tests, the substrates consisted of annealed carbon steel, and the thickness of the chrome deposits was of the order of 20 μm. All the test pieces were still intact after 100 hours whereas, in several cases, an electrolytic hard chrome coating showed initial corrosion pits after only a few tens of hours exposure.

3.6 Friction tests

Friction tests were carried out using a tribometer HEF device, the principles of which are illustrated in figure 11. The surface contact was of a cylinder/plane type with the test pieces consisting of small parallelepiped sections of plate having a slow backwards and forwards motion. These were pressed with differing forces on a rotating ring under the following conditions :

- ring material : high speed M2 steel
- test piece substrates consisted of 1038 steel with 20 μm coatings
- test environment : water
- load applied between rotating ring and oscillating plate increased by increments of 10 kgf every 10 minutes
- sliding speed was 0.55 m/sec.

The results of these tests are summarised in table 1.

It can be seen that Cr,C and Cr,N coatings possess superior characteristics to those of pure Cr and electrolytic hard chrome. Figures 12 and 13 show the appearance after testing of two small plates coated respectively by the electrolytic process and by the Cr,C sputtering process. In the first case (electrolytic hard chrome) cracking has occurred over the whole surface, whereas for the sputter deposited coating only a few score marks parallel to the direction of rubbing can be seen.

3.7 Rotating and bending fatigue tests

These tests were carried out on Moore type test pieces made from 4135 steel quenched and tempered at 570° C and with 3 different surface treatments. The values of the fatigue limits obtained were as follows :

- test piece no coating : 58.3 kgf/mm^2
- test piece coated with electrolytic hard chrome : 51.3 kgf/mm^2
- test piece coated by Cr,C: 54.80 kgf/mm^2

Figure 14 shows the Wöhler curves for these 3 differently coated test pieces. It can be seen that electrolytic hard chrome reduces the fatigue limit of the 4135 steel test pieces by 12 % compa-

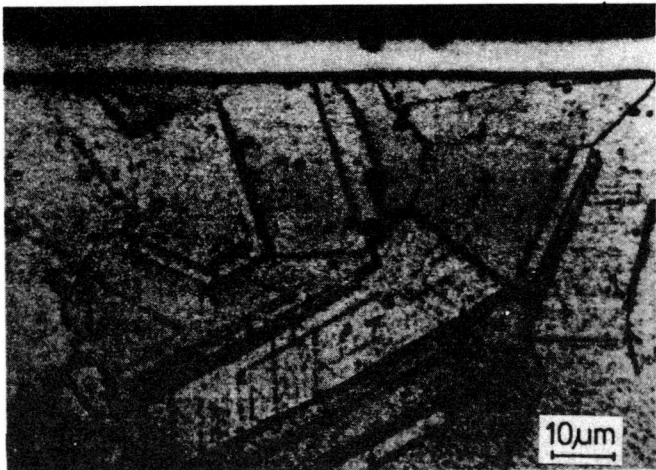

5 – Optical micrograph of pure chromium deposit on stainless steel

8 – SEM fractograph of a Cr,C film deposited at $T_s = 300°$ C and $P_{CH_4} = 3 \cdot 10^{-4}$ torr

6 – Scanning electron micrograph of a fractured region of a pure chromium deposit

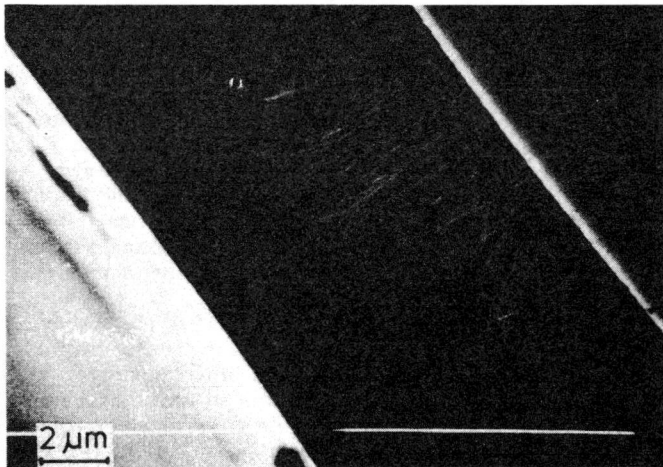

9 – SEM fractograph of a Cr,C film deposited at $T_s = 300°$ C and $P_{CH_4} = 8 \cdot 10^{-4}$ torr

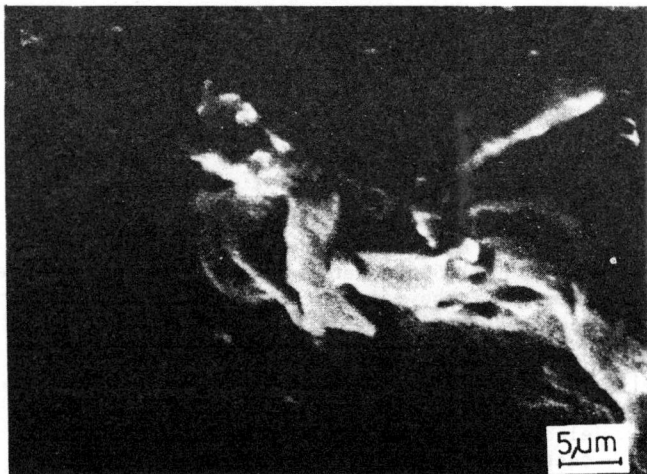

7 – SEM fractograph of a Cr,N film deposited at substrate temperature $T_s = 300°$ C and $P_{N_2} = 2 \cdot 10^{-4}$ torr

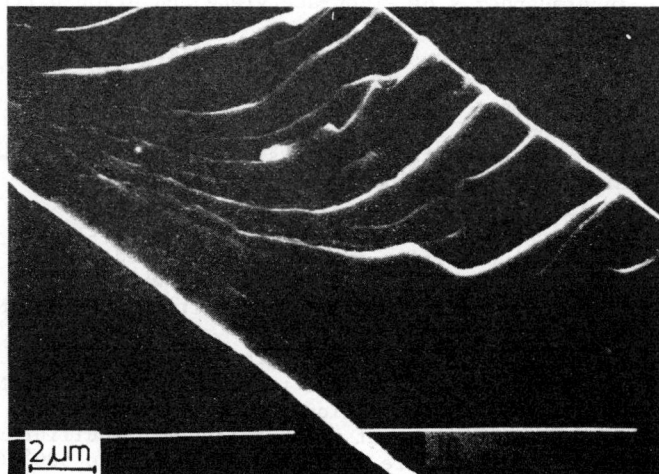

10– SEM fractograph of a Cr,C film deposited at $T_s = 300°$ C and $P_{CH_4} = 10^{-3}$ torr

1038 steel   environment: water

M2 steel   Load applied: increased by increments of 10 kgf every 10 mn.

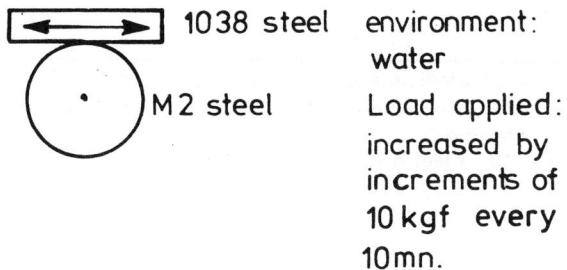

11- Friction test principles

Table 1_ Results of friction test.

Load kgf	10	20	30	40	50	60
Pure Cr	f=0.5	seizing				
electro-lytic hard Cr	f=0.3	f=0.27	f=0.27	seizing		
Cr,C	f=0.23	f=0,23	f=0.23	f=0.22	seizing	
Cr,N	f=0.23	f=023	f=0.23	f=0.23	f=0.23	seizing

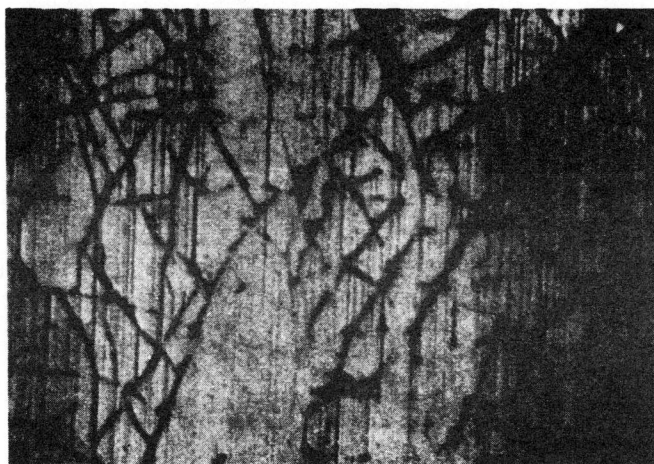

12- Electrolytic hard chrome deposit. Appearance after friction test

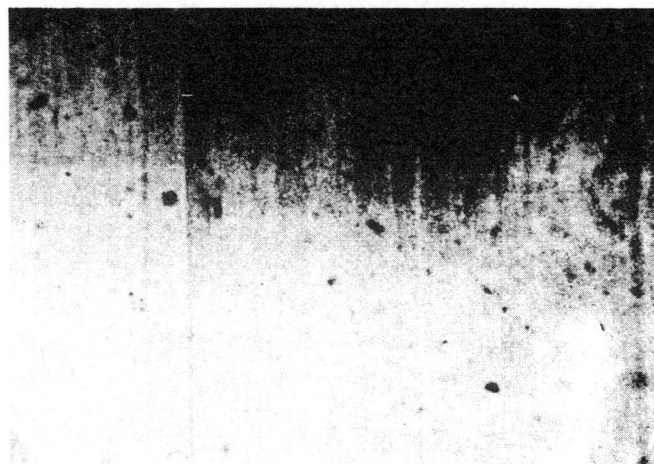

13- Cr,C sputtered deposit. Appearance after friction test

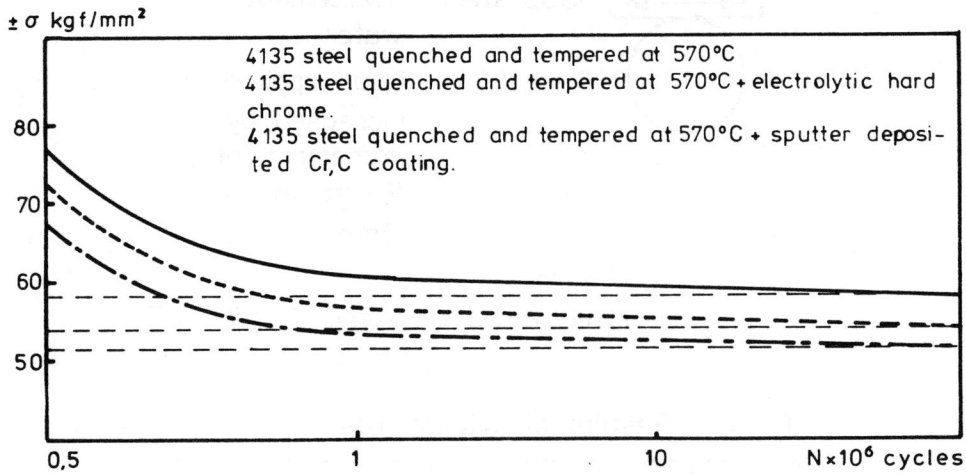

14- Wöhler curves for the three differently coated
test pieces

red to a reduction of only 6% for the Cr,C coating.
These differences can be explained by :
- hydrogen embrittlement of electrolytically coated
4135 steel and by the presence of many cracks
which act as notch type failure initiators
- Although Cr,C coatings do not exhibit these two
problems, the substrate preparation conditions
for these coatings (chemical cleaning, ionic
etching) and also their high hardness can reduce
the mechanical characteristics of the steel.

4. CONCLUSION

Thus in the present investigation (ref. 2), it
appears that appropriate doping of sputter deposited Cr coatings with C or N yields deposits with
superior mechanical properties and which in some
applications could replace electrolytic hard
chrome deposits.

Further investigations and analyses are pre-
sently being carried out to optimise these coating
properties.

ACKNOWLEDGMENTS

This research programme was partially supported by DRET.
We also wish to thank G. Uny and L. Lombard
for help with X-ray and SEM analyses, and J.P.
Chabert, J. Chevallier, J. Ernoult, J.P. Terrat for
experimental support and stimulating discussion.

REFERENCES

1 - S.Komiya, S. Ono, N.Umezu and T.Narusawa
Thin Solid films 45 (1977), 433-435.
2 - A.Aubert, J.Chevallier, A.Gaucher, J.P.Terrat
French Patent 81 17040, 3/9/81.

W D MÜNZ, D HOFMANN, and K HARTIG

High rate sputter process for the formation of hard, friction reducing TiN coatings on tools

SYNOPSIS

TiN coating for machine tools can reduce wear and friction and improve service life. A high rate sputter process is described. Measurements and tests used to optimize the coating process are given. For coating contoured substrates, a production system with two high-rate cathodes has been used.

THE AUTHORS

are in the Department of Research and Development for Coating, Leybold-Heraeus GmbH, Hanau, FRG.

INTRODUCTION

TiN and TiC coatings provide hard tips and cutting edges for machine tools as well as reduced wear and friction. Remarkable improvements in service life of tool and equipment are gained. The reduced friction and less friction welding at the cutting edges result in better quality of the surfaces produced.

All these improvements are well known from CVD-coated hard metal tools and cutting bits (Refs. 1 and 2).

To transfer these improvements in quality by TiN or TiC coatings from hard metal tools to the broad field of applications for high speed steel tools it is necessary to reduce the coating temperature below 550°C. Since attempts to develop low temperature CVD processes have not been successful so far, the PVD processes, mainly ion plating, have been used with remarkable success in multiplying the lifetime of high speed tools (Refs. 3—7). Up to now a handicap to extensive use of TiN coatings has been the low productivity and high cost of the relatively small coating units mostly based on laboratory coating units. There is limited use for a coating which is 4 times as good as an uncoated tool but at a price factor of 3.

We want to show that well controlled high rate sputtering of TiN can be used to coat tools at temperatures up to 600°C with good uniformity of thickness up to 3 μm around the tool in short cycle times.

This step forward in industrial PVD tool coating technique has been achieved by making use of experience in other fields of big industrial coaters using the high rate sputter process such as architectural glass coaters, where cycle times of some few minutes have already been achieved for 3 x 6 m panes with film tolerances known for optical applications in the percent range (Refs. 8—10). In the case of these coaters the step from slow batch systems to fast inline machines has been successfully accomplished. Due to the need for close tolerances, control of the high rate sputter process had to be developed to a high standard.

Before the new sputtering process is described in detail it should be mentioned that the high rate sputtering process is a typical 'cold' deposition process, first applied to the coating of semiconductor devices (Refs. 11—13) with Al, AlSi and AlSiCu alloys where substrate temperatures of 230°C must not be exceeded.

DOUBLE CATHODE CONFIGURATION

The ion plating process usually operates in a temperature range between 300 and 550°C. This is essential to achieve hard TiN films with the proper crystalline structure. Therefore special provisions must be made to convert the 'cold' high rate sputtering process into a 'hot' mode deposition. Simulating the ion plating process, the high rate sputtering process has to be run in such a way that the substrates to be coated can be biased so that, with sufficient ion bombardment, a temperature increase into the desired range is guaranteed.

The method of managing this has already been fully described (in Refs. 14 and 15) using a double diode configuration. Here the basic idea should be reviewed. The set-up of this arrangement is shown schematically in Fig. 1. The part to be coated is fitted between the high rate cathodes. The 'cold' and 'hot' deposition mode is characterized by the special extension of the confined plasma in front of the target area. The extension itself depends strongly on the power dissipated at the cathode, the magnetic field strength and the gas pressure. Figure 2 demonstrates the possibility of drawing an ion current from a high rate cathode as a function of the distance between substrate and target for two different values of the magnetic

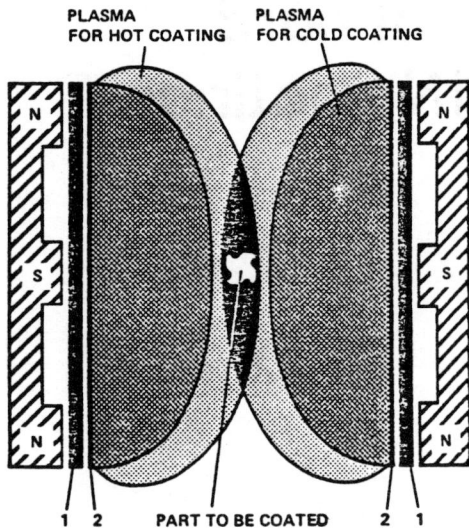

PRINCIPLE OF CONFIGURATION: ZONES FOR HOT AND COLD COATING

1 - CATHODES
2 - TARGETS

1 Double cathode configuration

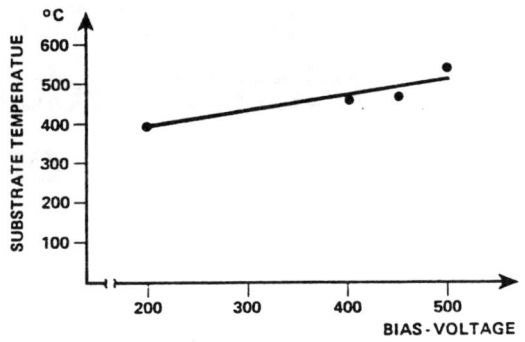

3 Substrate temperature of dummy measured using a Pt-100 resistance as a function of bias voltage

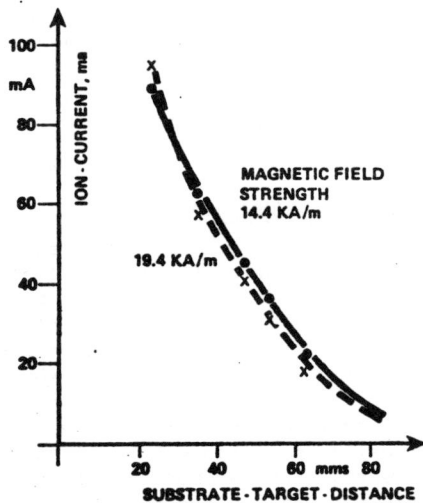

2 Bias current drawn at constant voltage (—40 V) as a function of the substrate/target distance

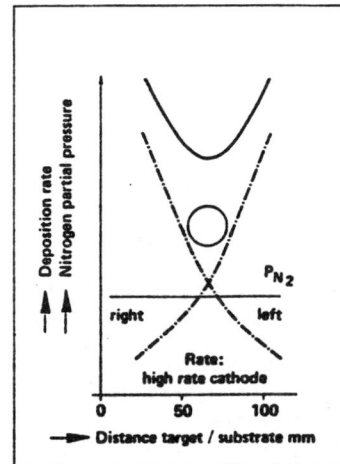

4 Schematic diagram of deposition rate and nitrogen partial pressure of two cathodes mounted opposite each other

field. This figure shows that the mechanism described in Fig. 1 is only effective if the two cathodes are mounted close enough together. As a result of such provisions Fig. 3 represents the temperature of a drill-shaped dummy of 10 mm diameter as a function of bias voltage. These results have been achieved with industrial-type 23 cm wide high rate cathodes. The working pressure was in the range of 2×10^{-2} mbar, the discharge voltage —500 V, and power dissipated at the cathode 9 W/cm^2.

The main advantage of the double diode configuration is its positive effect with respect to uniform rate distribution. Figure 4 shows schematically the decrease of the sputter rate from the 'left' and 'right' cathodes. Superimposing both curves, one recognizes a rather homogenous region between the two cathodes. In this limited range the deposition process has to occur. If the dimensions of the substrate do not exceed these geometrical dimensions, the substrates need not be rotated during the coating process. The stoichiometry of the films is still assured, due to the uniform rate distribution and the uniform distribution of the reactive gas pressure.

EXPERIMENTAL RESULTS

For the experiments the industrial coating plant described below was used. The hardness of the TiN layers was measured on flat polished samples of stainless steel. The thickness of TiN film was typically 4—5 μm. For the measurement a Leitz microhardness test unit was used. The hardness depends strongly on the partial pressure of nitrogen during coating as described in Refs. 14 and 15. Typical results range between HV_{10} = 2200 and HV_{10} = 3000.

The structure of the deposited films was studied by SEM of cross-sections of 10 mm dia. drills. Figures 5, 6 and 7 reveal the influence of the bias voltage and the resulting substrate temperature (compare Fig. 3). It is clearly shown that higher bias voltages (—500 V) result in more densely packed films while columnar structure dominates at lower bias (—200 V to —400 V). TiN films grown by ion plating show a structure similar to those in Fig. 6.

Although a bias voltage in the range of —500 V has almost the same value as the sputtering voltage, the observed overall reduction in film thickness was only 15% compared to a bias voltage of —200 V.

For use on tools, the uniformity of TiN layer thickness is of importance. The double cathode high rate sputter process gives (even without rotation of the substrate) a highly uniform thickness distribution. Figure 8 shows the thickness of the TiN film on a 5 x 5 cm polished steel substrate which has been coated stationary in a position perpendicular to the cathodes. On top of the plate where no shadowing took place the thickness is a constant 3.3 μm. On the bottom side shadowing by the plate holder at 1 cm distance resulted in a decrease of thickness to 1.5 μm at the centre of the plate. On both sides the structure of the film is most dense and smooth towards the ends of the test substrate. Less reflectivity has been observed in the centre region. Preliminary results show no remarkable deviations in Ti/N ratio and hardness across the test substrate.

Figure 9 shows the measured thickness distribution around a 12 mm tool-shaped sample, coated stationary. The mean thickness of 2.16 μm shows a ± 0.25 μm variation around the tool.

The stability parameter of the reactive process was observed by Auger electron depth profiles. They show excellent stability during a run of the machine and allow adjustment of the desired TiN ratio; Figs. 10 and 11 show two examples. The atomic concentrations are shown over the ion charge used to remove the layer for depth profiling. The vacuum conditions of the machine are easily detected by the O and C concentrations.

TECHNICAL REALIZATION OF
PRODUCTION PLANT

One of the most important advantages of sputtering is its obvious compatibility with in-line process realization. This again is a prior condition for reducing the cycle time of the comparatively long coating time necessary for the deposition of TiN films in the thickness range 2—5 μm. An apparatus which is capable of a cycle time of between 40 and 60 minutes is outlined in Fig. 13. The initial heat-up of the substrates to a temperature level between 400 and 500°C occurs in the input vacuum interlock. This chamber is equipped with a set of mechanical pumps including a Roots blower and a cold trap. The chamber is separated by a valve from the actual process chamber where sputter etching and film deposition occurs. The volume of the process chamber is approximately 3000 l. Two pairs of cathodes, each cathode 1.6 m long and 23 cm wide, provide the coating of 100 to 200 drills, end mills, etc. mounted in a frame 1.2 x 0.6 m. The distance apart of the cathodes is adjustable to the size of the substrates up to 200 mm diameter. Small diameters up to 25 mm allow a static coating mode; if this diameter is exceeded the substrates have to be rotated. The substrate carrier frame oscillates during the sputtering period, thus increasing the uniformity of coating thickness.

The process chamber is equipped with turbo-molecular pumps which ensure the adequate and stable pumping speed essential for a reproducible reactive sputtering process. Gas composition is controlled by flow meters.

All the process steps, sputtering as well as the etching prior to sputtering, are run in the DC mode. The power supplies for the etching process are voltage-controlled whereas the sputtering process is current-controlled.

When the coating process is finished the substrate carrier frame moves into the exit vacuum interlock, which is also separated by a valve from the process chamber. Here the substrates are allowed to cool in a non-reactive atmosphere to at least 200°C. From there the frame leaves the apparatus via the exit valve. The pumping installation is similar to that of the input vacuum interlock and consists exclusively of mechanical pumps.

The actual productivity of the plant depends, of course, on the dimensions of the substrates, the loading density, and the coating parameters, so that no general prediction can be made on the final capability of the plant.

Finally it should be mentioned that all the process steps operate fully automatically.

CONCLUSIONS

The coating method described demonstrates the capability of the high rate sputtering technique for the formation of hard TiN films. The results reveal also that there is no relevant difference in the physical quality of high rate sputtered films compared to the published data on ion plated films.

This coating technique allows a broad variation of coating parameters. To achieve the greatest

5 Structure of TiN film formed with bias voltage
—200 V and resulting temperature 400°C
(thickness: 300 µm)

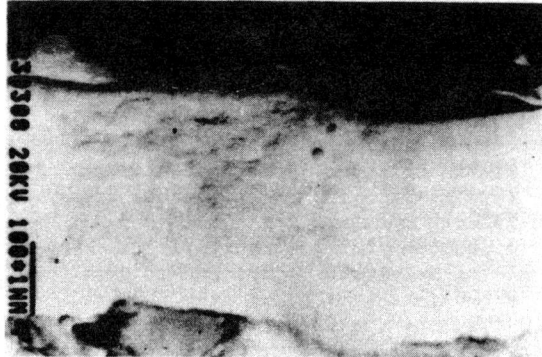

7 Structure of TiN film formed with bias voltage
—500 V and resulting temperature 530°C

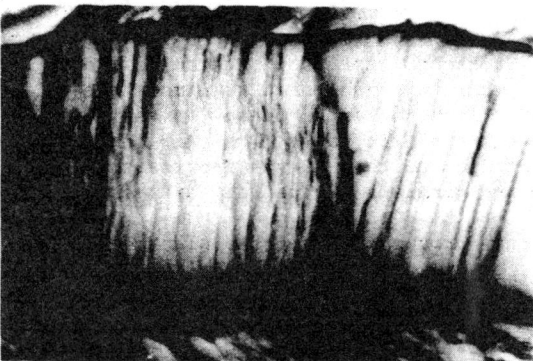

6 Structure of TiN film formed with bias voltage
—400 V and resulting temperature 450°C
(thickness: 2.89 µm)

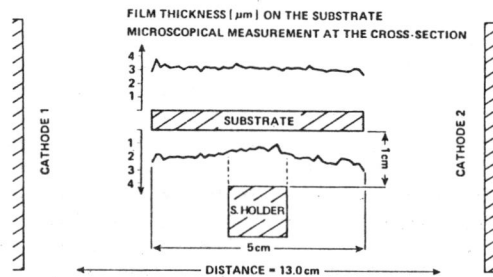

8 Thickness distribution of 3 µm thick TiN film on
5 x 5 cm substrate mounted perpendicular to
cathodes

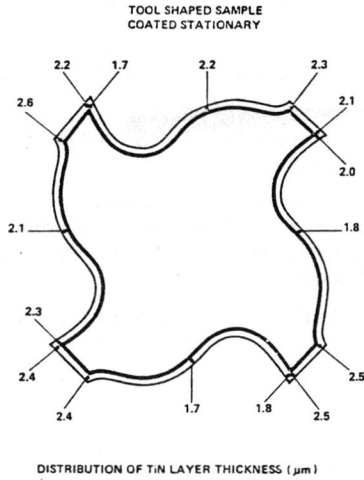

9 Thickness distribution around tool not rotated during film deposition

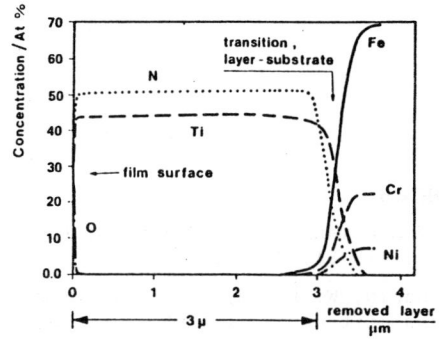

11 Auger depth profile of TiN film over whole thickness

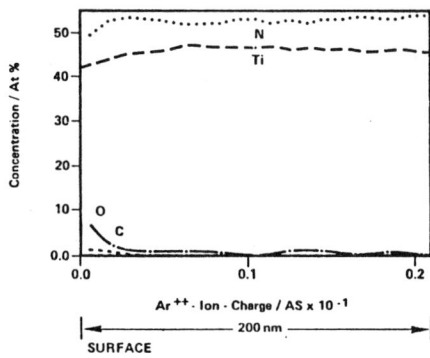

10 Auger depth profile of TiN film in vicinity of surface

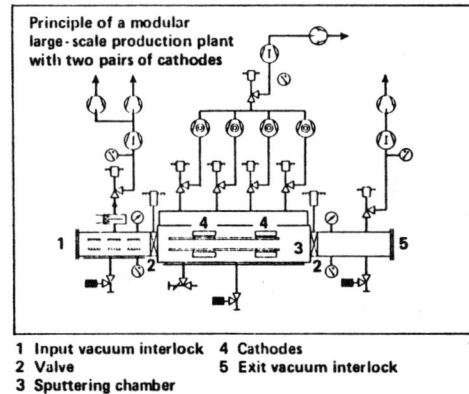

12 Schematic diagram of large-scale production plant

advantages of coating, however, it is certainly necessary for the coating process to be adapted carefully to the conditions of each tool. This needs optimization work and special preparation of the substrates to be coated.

REFERENCES

1 H. E. Hintermann, A. G. Perry, E. Horvath: Wear, 47 (1978), 407-415

2 K. H. Habig, W. Evers, H. E. Hintermann: 7. Werkstofftechnik 11 (1980), S. 182

3 D. M. Mattox: J.Vac.Sci. Technol., 10 (1973), 47

4 D. G. Teer: Proc.Int.Conf. Ion Plating and Allied Techniques (IPAT), Edinburgh 1977, 13-31

5 R. F. Bunshah: Thin Solid Films, 80 (1981), 255

6 I. Barany, G. Kienel: Vakuumtechnik, 28 (1978), Heft 6, S. 168

7 VDI-Nachrichten Nr. 45, 7 Nov. 1980, S. 14

8 G. Kienel, B. Meyer, W. D. Münz: Vakuumtechnik 8/1981, 236

9 W. D. Dachselt, W. D. Münz, M. Scherer: SPIE's Los Angeles Technical Symposium and Instrument Exhibit, Jan. 25-29, 1982

10 W. D. Münz, S. R. Reineck: SPIE's Los Angeles Technical Symposium and Instrument Exhibit, Jan. 25-29, 1982

11 L. D. Hartsough, D. R. Denison: Solid State Technology, Dec. 1979, 66

12 J. F. Smith and W. Chass: paper presented at International Conference on Metallurgical Coating and Process Technology, San Diego

13 S. R. Reineck, W. D. Münz: Elektronik Produktion and Prüftechnik, Jan. 1982, 77

14 W. D. Münz, G. Hessberger: Vakuumtechnik, 30 (1981), 78

15 W. D. Münz, G. Hessberger: Industrial Research and Development, Sept. 1981

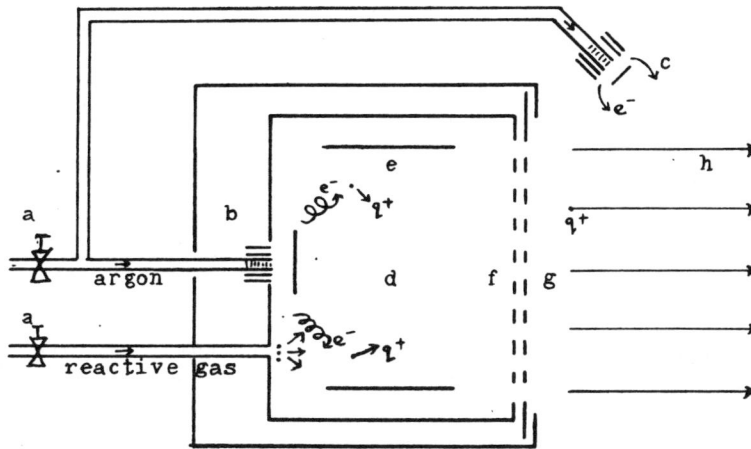

(a) gas control valve
(b) source hollow cathode (+500V)
(c) neutraliser hollow cathode
(d) discharge chamber

(e) anode (+550V)
(f) screen grid (+500V)
(g) accelerator grid (-200V)
(h) ion beam (+500eV)

No attempt has been made to show the multipole magnets which increase the path length of the electrons and hence their ionisation efficiency.

4   Schematic diagram of an electron-
    bombardment ion source

5   A dual ion-beam system in operation

2. THE PRINCIPLES OF ION BEAM MACHINING

Argon ion beam milling

The usual source of ions is argon because of its inertness and its low cost. An argon ion striking the surface to be machined will have typically a yield of between 0.1 to 10 atoms removed per incident ion. This yield depends upon the angle of incidence and the energy of the ion. The yield increases with increasing angle of incidence θ for almost all surface materials apart from gold: so much so that the net milling rate of most materials is a maximum at an angle of incidence of θ = 40 to 60 degrees, despite the cos θ reduction in effective current density at the surface. (See figure 1b)

The yield also increases with increasing ion energy[8,9]. There is a threshold of 100eV. Yield increases rapidly to about 1keV and at an ever-decreasing rate thereafter. For most materials the value of yield/ion energy maximises at between 300 and 500eV. Thus surfaces are machined with the minimum heating in this energy range. Since the available current density varies as voltage to the 3/2 power (see equation 2 below), 500eV is a suitable value. At this energy, surface damage is minimal - germanium, for example, is damaged to a depth of 11 atomic layers[10].

In many applications it is required to mill features through a mask and to ensure that the walls are vertical and that the floor is flat. Vertical walls are not obtained at normal ion incidence due to two effects: trenching and re-deposition[11] (see figure 1a). The former effect is the production of trenches at the base of feature walls due to the reflection of ions off these walls at grazing incidence and the resulting enhanced milling rate at the base. Redeposited atoms on the walls will usually be structurally satisfactory, but may also change the profile in an undesired manner.

Fortunately, both these effects may usually be minimised by inclining the surface to the ion beam, and rotating the surface about its normal axis. By fine adjustments to the angle of incidence θ the walls can usually be adjusted to within a few degrees of vertical.

The nature of the milled surface can be very different according to the composition of the material and the milling conditions used. An ion beam can be used to texture the surface, covering it with cones or ridges with dimensions of the order of 1 micrometer[12,13]. Alternatively the beam can smooth the surface to a sub-micron roughness[14] or, if the surface is already smooth, can machine it away without increasing the roughness significantly - typically 100nm can be removed with a rms surface deviation of 1.5nm changing by less than $\pm$1nm[15].

It is not possible to predict the exact nature of the surface that will be produced, as much depends on the crystalline structure of the surface and the relative milling rates of the crystallographic planes exposed to the beam. Little of such data is yet documented. However, certain guidelines can be laid down: the surface will become textured if it is stationary in the beam, if it is contaminated with a "seed" material such as tantalum[12], if the surface temperature rises high enough to produce surface atom mobility.

Conversely, rotation of the surface under an ion beam with an angle of incidence of 40-80 degrees, clean surfaces and low surface temperatures all have a tendency to produce smooth surfaces.

Disadvantages of argon milling

Four main disadvantages of ion beam milling must be acknowledged:

1. Cost Because this is a high vacuum process, equipment costs are indeed high, especially if automated airlock operation is envisaged. However, the same is true of other techniques, such as electron-beam welding, which are firmly-established production processes.

2. Milling rate Much of the energy of the ion beam is dumped as unwanted heat in the specimen. This limits the rate at which one can mill through masks of organic photoresist to a material removal rate of typically 1 nm/s at normal ion incidence, and 2 nm/s at 50 degrees incidence (see figure 2). However, milling rates one to two orders of magnitude higher can be managed if the mask and substrate are both unsusceptible to heat. We are also finding that a promising technique: Reactive Ion Beam Milling, as described below, can increase the milling rate very substantially - typically by a factor of ten.

3. Component shape Uniform ion milling operations can only be performed on components with simple shapes: planar surfaces, cylinders and cones. Spherical surfaces, even, are difficult.

4. Selectivity It is common to find that the ion beam mills the mask almost as fast as the material being machined. For example: aluminium 1.2nm/s AZ1350 photoresist 0.5nm/s.

Reactive Ion Beam Machining

In reactive ion beam machining (RIBM) a flux of reactive species is directed at the specimen, instead of the flux of inert argon ions.

Typically, the source is fed with a gas such as CF_4, C_2F_8, CH_4 or O_2.[4] A whole range of ionised species such as CF_4^+, CF_3^+ and F^+ is produced in the source. These ions are neutralised by Auger or resonance processes as they approach to within 1nm from the surface. Thus CF_4^+ interacts with the surface as the inert CF_4, but CF_3^+ becomes the highly reactive free radical CF_3. If the beam energy is sufficiently low ($\leqslant$100eV) the free radical will have sufficient residence time to react with the surface.

For effective reactive milling, the free radical chosen will not react with the mask material, but will react with the surface to form products which are either volatile or are easily milled by the kinetic energy of the ions. A typical process is the milling of copper aluminium alloys using CCl_4 through an organic photoresist mask.

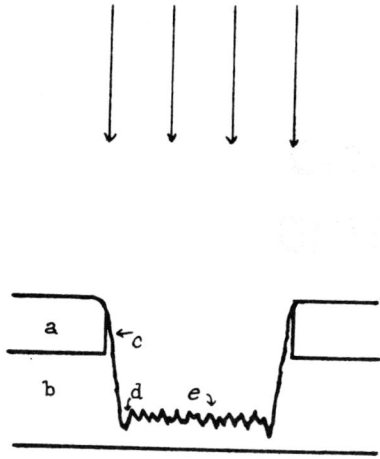

1a Ion Milling at normal incidence
   (a) mask (b) material (c) redeposition
   (d) trenching (e) textured surface

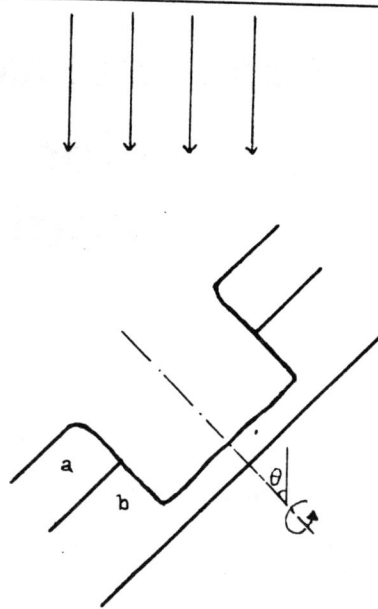

1b Ion Milling at an angle θ

Target Material	nm/s
C	0.07
Al	1.2
Si	0.62
Cr	0.83
Mn	1.5
Bi	15
Fe	0.67
Co	0.80
Ni	0.83
Cu	1.42
Ag	2.5
Ta	0.63
W	0.56
Au	2.4
Pb	5.2
Sn	2.0
SiC (0001)	0.52
SiO_2	0.67
Al_2O_3	0.21
Fe_2O_3	0.78
AZ 1350 (photoresist)	0.50
PMMA (photoresist)	0.93

2 Some milling rates for 500eV argon ions at $1mA/cm^2$ and normal incidence.[16]

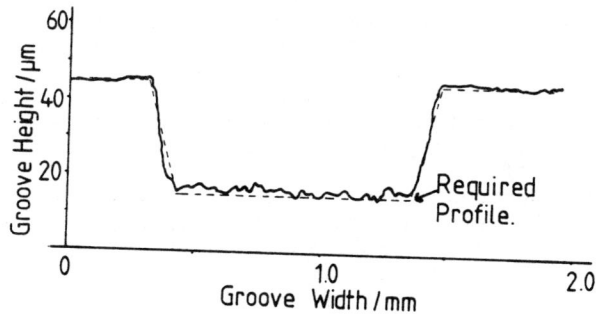

3 Profile of groove machined in a refractory metal surface

T W JOLLY and R CLAMPITT

Ion milling and coating systems

SYNOPSIS

Low energy ion beams find increasing application in machining of component surfaces. The technology is finding diverse rôles in engineering, medical and other industries where materials can be textured and micromachined to impart unique catalytic, biological, adhesive or wear-resistant properties. Dedicated processing systems have been designed in which dual mill and coat processing of components can be achieved in a one-step operation. We will outline these emerging applications of low energy ion beam processing and describe a variety of typical systems available to users.

THE AUTHORS

R. Clampitt is managing director of Oxford Applied Research, Crawley Mill, Witney, Oxon., England. T. W. Jolly is a senior research scientist with the company.

1. THE ORIGINS OF ION BEAM MILLING

It has been recognised for over forty years that energetic ions have unique properties for cleaning and milling surfaces[1]. These properties arise from the fact that if a surface in vacuo is bombarded with energetic ions (or atoms - as is shown below there is no significant difference) then atoms are sputtered from the surface by momentum transfer. They are broken free singly or in very small clusters. By way of contrast, in laser and electron beam processes the surface is heated until droplets of material are boiled off.

The semiconductor industry realised about six years ago that ion beams could solve production problems that were becoming very difficult[2]. The processes in question involved the machining of features a few micrometers across using a mask produced photographically to define the pattern. Until then, wet chemical etching or plasma processing had been used. Ion beams offered a number of substantial advantages:

1. It is possible to machine any material - although graphite and boron nitride are difficult to machine due to their low milling rates.

2. The size of feature which can be machined is exceedingly small: mask-making techniques now allow sub-micron dimensions. Features of less than 100nm have been achieved using fine scanning ion beams[3] and 10nm should be possible.

3. The angle of incidence of the ion beam is fully controllable. This gives full control over the slope of machined groove walls and over the final surface finish of the machined area.

4. The removed material does not form dust or droplets, and it is possible to eliminate any re-deposition of such particles.

5. The process parameters can be monitored to an accuracy of ±1% and a repeatability of ±0.25%. Thus suitable controllers can give a process repeatability of ±1%.

6. The process is accurately scaleable. Ion sources giving beams 80cms across have the same properties as 2.5cm sources.

7. Because it is a high vacuum process, extremely sophisticated vacuum techniques such as electron microscopy and secondary ion mass spectrometry (SIMS)[4] can be deployed both for process control and for quality assurance.

By a convenient coincidence, the one suitable type of ion source, the electron-bombardment source or "Kaufman thruster", has been developed continuously since 1960[5,6]. For the first decade it was developed as a relatively low-thrust motor for use on satellites for attitude and orbit adjustment. Sources were developed which produce substantial currents of mercury or xenon ions and which could run for several thousand hours without attention[7]. For ion milling, the sources were modified to run on argon, and to give large diameter uniform parallel beams[2,6]. An unfortunate side-effect of the switch to argon was a reduction, in earlier milling sources, of the running time of the source between maintenance to an unpredictable interval of between 10 and 20 hours. This was due to the failure of the tantalum or tungsten wire cathodes in these sources.

The beam is used at normal incidence. (Negligible trenching or redeposition occurs.) Features with depth-to-width ratios of greater than 10:1 have been achieved.

Ion beam sputter deposition

If a metal target is bombarded with argon ions, the sputtered products are largely single atoms with less than 1% ionised. According to the collision cascade model[17] of Sigmund[17] and its later developments[18], the sputtered material has an energy and angular distribution given by:

$$\Phi(E,\phi) = CE \cos \phi \: / \: (E + E_b)^{n+1} \qquad (1)$$

where c and n ($1 < n \leqslant 2$) are approximately constants of the ion and substrate material combination and E_b is the surface binding energy. The sputtered material is emitted with a $\cos \phi$ distribution about the normal of the target.

The sputtered material using 1500 eV Ar^+ ions typically has an energy peak in the range 10-30 eV, and a small percentage at over 100 eV[18]. These particle energies contribute to the success of ion beam sputter deposition.

Targets such as oxides and p.t.f.e. sputter a range of particles - both atoms and small clusters and these particles all contribute to the coating which is produced.

Ion beam sputter coating is being performed in figure 5, where the ion source at the lower left is sputtering material off the target in the centre onto the specimens on the large holder in the top left. Alumina was being deposited on an aluminium surface at 0.5nm/s with a uniformity of $\pm 2\%$ over 10cm[16].

The advantages of the process include:

1. It is performed in a high vacuum, and the surfaces can be atomically clean if a pre-cleaning ion gun is used.

2. Little involuntary surface heating occurs.

3. The angle of deposition is easily varied.

4. Good adhesion results from the energy of the coating particle flux.

5. Certain materials - notably fluoro-polymers such as p.t.f.e. can, perhaps surprisingly, be coated very successfully onto metals[19].

Reactive sputtering is also a successful process - for example milling a silicon target with a nitrogen ion beam gives coatings of Si_3N_4.

Dual beam processing

A dual beam system of the type in figures 5, 8 and 9 introduces so many variables that the number of possible processes becomes legion. However, in principle the system has one source dedicated to sputtering the target (usually with argon ions) and a large second source directed at the surface being coated. This second source may be used for precleaning, for milling of the deposited coating or for simultaneous processes some of which are described below.

3. TYPICAL APPLICATIONS OF ION BEAM MILLING

Milling through masks

The ion milling of grooves in engineering components is a tested process[20], and it is especially valuable for the accurate production of shallow grooves: we show an example in figure 3.

Ion Beam Texturing

This process has many applications[13]. Some are of immediate interest: enhanced bonding of surfaces, such as fluoropolymers to metals, increased surface areas of capacitors, nucleate boiling heat transfer surfaces. Most applications require further development: surface treatment of medical implants (both hard and soft) for biocompatibility, non reflective coatings for solar cells[21], increased surface area for catalysts.

Ion Beam Cleaning

The cleaning of steel using electron beams in vacuo has been used widely for some years, and cleaning of semiconductors prior to metallisation using low pressure electrical discharges was until recently almost universal. Both processes, however, can damage the surface and tend to "re-cycle" the contamination back on to the surface. Ion beam cleaning, however, can produce atomically clean surfaces, and sources producing large rectangular beams (up to 5cm x 1.2m) are currently being retro-fitted to numbers of vacuum coating and metallisation systems (see figure 7).

Ion Beam Smoothing

This process has been developed for smoothing laser mirrors[14] and for adjusting the thickness of thin films and membranes without affecting the surface finish[15].

Ion Beam Sputter Deposition

Current non-semiconductor uses for ion beam sputter deposition include coatings of molybdenum, tungsten and platinum for aluminium casting dies[13], and multilayer dielectric reflective coatings for laser mirrors. P.t.f.e. coating of blades is also a production process. With the capability of depositing mixed materials such as "Teflon"-copper, many new applications remain to be discovered, and much work is being performed on lubrication coatings, anticorrosion oxide coatings, capacitor manufacture[22], the diamond-like coatings of Weissmantel[23] and so on.

Dual Beam Processing

The dual beam process in figure 5 involves the sputtering of alumina as already described. In addition, the source at the right is producing a 15cm diameter parallel beam of oxygen to clean the substrate before coating and to adjust the stoichiometry of the coating applied.

In another process, niobium coatings are bombarded with argon ions during formation to control the oxygen content[24], or alternatively with nitrogen ions to produce niobium nitride coatings.

The second beam has also been used to re-sputter the coating selectively off the surface as it is deposited. The net

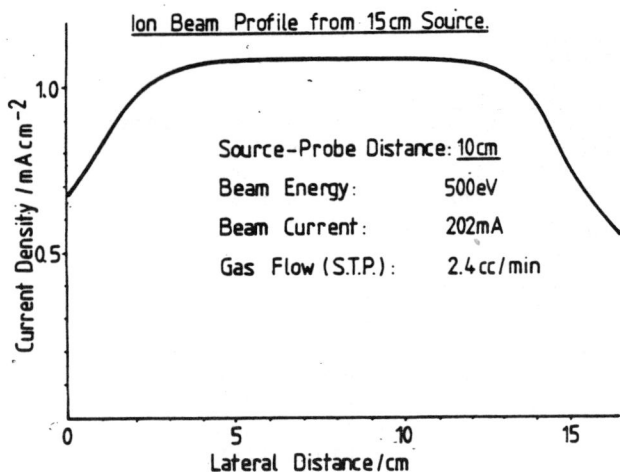

Ion Beam Profile from 15cm Source.

Source-Probe Distance: 10cm
Beam Energy: 500eV
Beam Current: 202mA
Gas Flow (S.T.P.): 2.4 cc/min

6   Ion beam profile from a 15cm source

7   A thermal evaporation coater fitted with a 40mA pre-cleaning ion source, visible at the rear of the chamber.

result is uniform coatings over complicated topographies[25].

CURRENT ION MACHINING EQUIPMENT

A schematic drawing of a modern electron-bombardment ion source is shown in figure 4. Electrons emitted by the source's hollow cathode ionise the gas in the discharge chamber. The ions are then extracted efficiently by the double grid arrangement at the front of the source. The ion current which can be extracted through each aperture is given by Child's Law as[26]

$$I = \pi \epsilon_o /9 \quad V^{\frac{3}{2}}(2q/m)^{\frac{1}{2}}(d/\ell_e)^2 \qquad (2)$$

where ϵ_o is the dielectric permittivity, V is the potential difference between the grids, q and m are the ion's charge and mass, d is the hole diameter in the grids. ℓ_e is the effective grid separation, and is given by:

$$\ell_e = (\ell^2 + d^2/4)^{\frac{1}{2}} \qquad (3)$$

The maximum currents are thus achieved by minimum grid separation, maximum numbers of holes and minimum hole size. Electrons are emitted from an additional hollow cathode into the ion beam to ensure that the surface being machined does not develop an electrostatic charge.

2.5cm beam and 15cm beam sources are shown in operation in figure 5. The beam profile of this latter source is shown in figure 6. As mentioned above, much larger sources are available.

The specimen holder is shown in figure 8. The water cooled platen rotates continuously, and the angle of tilt can be adjusted from 0° to 90°.

Source controllers are visible at the top of figure 9. They are computer-controlled and employ switching-mode power supplies. They are fully automatic in operation - the user dials in the beam energy and current required, and all power supplies are then controlled automatically.

CONCLUSIONS

It is likely that major applications for ion beam processing remain to be invented now that the process exists. Although most potential exists in the processing of high-cost miniature components, it is likely that the general use of larger equipment, and the development of faster processing by such means as reactive ions, will make the process more generally economical.

REFERENCES

1. See the review by Farkas A. and Melville H.W. Experimental methods in gas reactions, Macmillan, London, 1939
2. Bollinger L.D. "Ion milling for semiconductor production processes" Solid State Technology vol. 20 no. 4 pp 66-70 (1977)

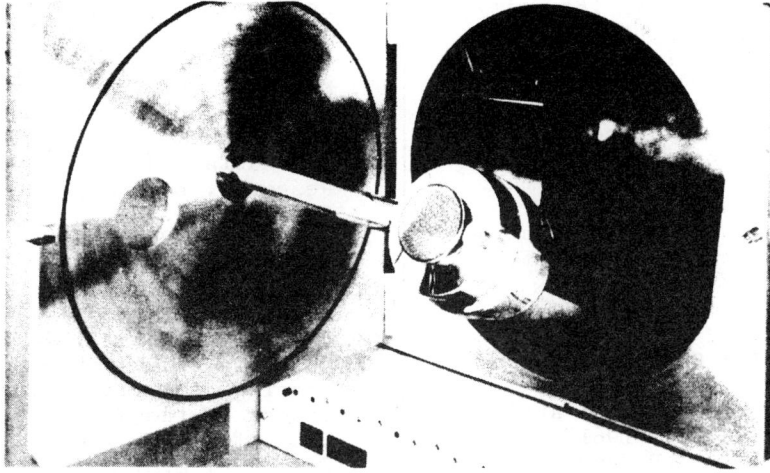

8   A water cooled specimen holder, with
    platen rotation and adjustable angle
    of tilt.

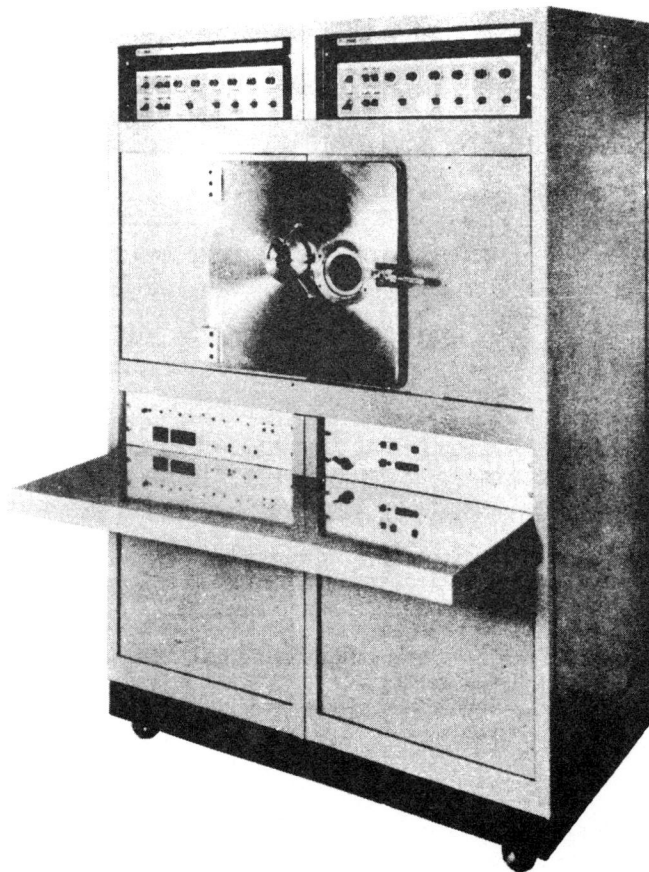

9   A dual beam milling system. Note the
    two source controllers at the top of
    the console.

3. Seliger R.L. et al. "A high intensity scanning ion probe with submicrometer spot size" Appl. Phys. Lett. vol. 34 no. 5 pp 310-312 (1979)

4. Clampitt R., Jolly T.W. and Reader P.D "Reactive Ion Etching and free radicals Proceedings of the technical programme Semiconductor '81 International, pub. Cahners, England (1982)

5. Kaufman H.R. and Reader P.D. "Electrostatic propulsion" American Rocket Society Paper No. 1374-60 (1960)

6. Reader P.D. and Kaufman H.R. J. Vac. Sci. Technol. Vol. 12 p 1344 (1975)

7. Azara H. et al. AIAA Paper no. 81-662 (1981)

8. Laegreid N. and Wehner G.K. J. Appl. Phys. Vol. 32 pp 365-369 (1961)

9. Southern A.L., Willis W.R. and Robinson M.T. J. Appl. Phys. Vol. 34 pp 153-163 (1963)

10. MacDonald R.J. and Haneman D. J. Appl Phys. Vol. 37 pp 1609-1613 (1966)

11. Chapman R.E. "Redeposition - a factor in ion-beam etching topography" J. Mater. Sci. Vol. 12 pp 1125-1133 (1977)

12. Hudson W.D. "Ion beam texturing" J. Vac Sci. Technol. Vol. 14 pp 286-289 (1977)

13. Banks B.A. NASA Technical Memorandum 81721 (1981)

14. Hoffman R.A., Lange W.J. and Choyke W. Applied Optics Vol. 14 pp 1803-1807 (1975)

15. McNeill J.R. "Ion Beam applications for precision infra-red optics" J. Vac Sci. Technol. Vol. 20 pp 324-326 (1982)

16. Data obtained by Oxford Applied Research, England, and Ion Tech Inc, Colorado, U.S.A.

17. Sigmund P. Phys. Rev. Vol. 184 p 383 (1969)

18. Krauss A.R. and Gruen D.M. Appl. Phys. Vol. 14 pp 89-97 (1977)

19. Banks B.A. et al NASA TM-78888 (1978)

20. Butler A. "Generation of hydrodynamic bearing grooving by ion machining" Trans. of the ASTME, J. of Lubrication Technology 334 (1975)

21. Rossnagel S.M. and Robinson R.S. J. Vac Sci. Technol. Vol. 20 pp 336-337 (1982)

22. Hudson W.R. Robson R.R. and Sovey J.S. NASA TMX 3517 (1976)

23. Mirtich M.J. AIAA Paper 81-671 (1981)

24. Cuomo J.J. et al. J. Vac. Sci. Technol Vol. 20 pp 349-354 (1982)

25. Harper J.M.E. Proto G.R. and Hoh P.D. Engineering Technology 9.9.80 (1980)

26. Aston G., Kaufman H.R. and Wilbur P.J. AIAA J. Vol. 16 p 516 (1978)

N J ARCHER

Plasma assisted chemical vapour deposition

THE AUTHOR

is with Archer Technicoat Ltd, High Wycombe,
Buckinghamshire.

SYNOPSIS

Plasma Assisted Chemical Vapour Deposition
(PACVD) is being developed to provide coatings
at temperatures much lower than is possible by
the corresponding Chemical Vapour Deposition
(CVD) method. The temperature of deposition
is critical to many applications where the
properties of the substrate are impaired by
the often high temperatures (700 - 1400°C)
demanded by CVD. PACVD opens up the possibility
of obtaining the desirable properties of high
temperature coatings without damaging the
substrates. The technique of PACVD is discussed,
and the applications which it has found so far
are described.

The Desire for Low-Temperature Coatings

Chemical Vapour Deposition (CVD) has become
established as an important method of obtaining
overlay coatings for both electronic and mechanical
applications[1]. The best known mechanical application
is the coating of sintered carbide cutting tips
with carbides, nitrides and oxides[2]. The indexable
sintered carbide tool tip is ideal for CVD coating;
neither the cobalt matrix nor the carbide phases
are altered significantly by temperatures in the
range 850-1050°C which are used in CVD, and there
is no need for subsequent processing. Steel tools,
which are far more numerous than carbide tools,
can also be coated by conventional high temperature
CVD. The high temperature causes a small amount of
interdiffusion which assists bonding between the
steel substrate and the coating, but it also softens
and stress-relieves the steel. The steel can be
rehardened in many instances, but often the amount
of distortion which has occurred during the whole
cycle is unacceptable on precision components.

With electronic devices the desire for low-
temperature coatings stems from the need to avoid
diffusion between substrate and coating. Discrete
epitaxial layers cannot be made thinner than the
interdiffusion zone between adjacent layers.

Methods of Reducing the Temperatures used in CVD

Before considering the ways in which deposition
temperatures in CVD may be reduced, it is necessary
to ask the reasons for the present choice of
temperature. The CVD method essentially consists
of introducing volatile reactants into a heated
chamber where a surface-activated reaction occurs
to give an adherent and coherent deposit. The
best known example of this is the reaction leading
to titanium carbide:-

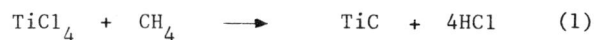

$$TiCl_4 + CH_4 \longrightarrow TiC + 4HCl \qquad (1)$$

Titanium tetrachloride and methane are both
volatile and stable at ambient temperatures. Moreover, they do not react at this temperature so they
may be mixed and introduced into a heated chamber
on a flow of inert carrier gas. Figure 1 shows the
variation in the $\triangle G$ for reaction (1) as a function
of temperature. This indicates that the formation of
TiC will not be expected at temperatures below 900°C
under equilibrium conditions. Under CVD conditions,
where the by-products are continuously being swept
away, the formation of TiC may be expected at somewhat lower temperatures[3], but there is a fundamental
thermodynamic limitation relating to the particular
choice of reagents. The situation is much the same
for the other CVD reactions listed in Table 1.

Within the context of CVD, there are two ways
of attempting to operate at lower temperatures;
(1) change the chemistry and look for reactions
which occur at lower temperatures, or (2) search
for methods of activating the known chemistry which
do not involve heating the substrate.

The search for lower temperature chemistry has
been successful in the electronics field where
metal organics have been used. Epitaxial gallium
arsenide can be formed by CVD from the appropriate
alkyls. Cadmium mercury telluride can also be
formed from the alkyls[4]. Thin layers of aluminium
can be obtained by the thermal decomposition of
tri-isobutyl aluminium. However, it has proved to
be more difficult to find corrosion and wear resistant layers which can be obtained at low temperatures.
Tungsten carbide can be formed from tungsten
hexafluoride and benzene at temperatures around
500°C, and this has been successful in some
applications. Coatings of silica have been obtained
from alkoxy silanes at 600°C (Table 2).

Table 1 Typical high temperature CVD reactions

Material Deposited	Chemical Reaction	Temperature of Substrate (^{o}C)	Application	Ref
BN	$BCl_3 + NH_3$	1000-2000	High Temperature Ceramic	8
TaN	$TaCl_5 + N_2$	800-1500	Resistor Film	9
Si_3N_4	$SiCl_4 + NH_3$	1100-1400	High Temperature Ceramic	10
TiN	$TiCl_4 + N_2 + H_2$	650-1700	Hard Coatings	2
B_4C	$BCl_3 + CH_2$	1300-1900	Armament	11
SiC	$CH_3SiCl_3 + H_2$	1000-1600	Oxidation Resistant Coatings	12
TiC	$TiCl_4 + CH_4$	800-1400	Hard Coatings	2
ZrC	$ZrCl_4 + CH_4 + H_2$	1050-1500	Diffusion Barrier	13

Table 2 Low temperature CVD reactions

Material Deposited	Chemical Reaction	Temperature of Substrate (^{o}C)	Application	Ref
Al	AlR_3 or $AlHR_2$ (thermal decomp)	200-600	Layers on Plastics and Electronic Devices	14
As	AsH_3 (thermal decomp)	230-300	GaAs Semiconductors	15
B	B_2H_6 (thermal decomp)	400-700	Fibres	16
Ge	GeH_4 (thermal decomp)	400-900	Semiconductors	17
Mo	$Mo(CO)_5 + H_2$	300-600	Thin Film Resistors	18
Ni	$Ni(CO)_4$ (thermal decomp)	180-200	Vapour forming	19
SiO_2	$SiH_6 + O_2$ and $Si(OC_2H_5)_4 + O_2$	300-430 600-900	Passive Encapsulant for electronic devices	20 21
TiO_2	$Ti(OC_3H_2)_4 + O_2$	450	Dielectric films	22
TiN	$Ti(NH_2)_4$ (thermal decomp)	300-500	Wear Resistant Layer	23
Mo_2C	$Mo(CO)_6$ (thermal decomp)	350-475		24
W_2C	$WF_6 + C_6H_6 + H_2$	400-900	Wear Resistant Layer	25

Two methods of activating CVD have emerged. One method uses a pulsed laser to create local hot spots on the surface of a substrate without heating its bulk[5]. This method is attractive when localised deposits are required, but it is not so suitable for the all-over coating of large numbers of components.

A limitation is that the laser frequency must not correspond to any of the molecular absorption frequencies of the reaction mixture. The other method of activation involves the use of a plasma zone around the components to be coated. The molecular excitation within the plasma zone induces reactions to occur without heating the substrate. This method has very general applicability and appears to be a serious contender for many applications when the substrate temperature must be limited.

The Use of Plasma in Conjunction with CVD

The well-known light emission from a plasma is caused by excited gas molecules returning to their ground status after being excited by some form of molecular collision. The partial ionisation of a gas and the subsequent acceleration of the ions and electrons in an applied electric field is the usual method of causing the high energy molecular collisions. A plasma may be established in any gas, but argon is particularly convenient. If a CVD reaction mixture is introduced into an argon plasma, some of the energy in the argon molecules will be transferred to the reactant molecules.

They will begin to react as if they were at a high thermal temperature corresponding to their state of excitation, although the state of excitation can easily be far in excess of any practical thermal temperature. The high temperature product then 'freezes out' on the substrate.

If the CVD reactants are too successful at transferring energy away from the argon plasma, it will be extinguished, and so a very careful balance must be maintained between CVD reactant concentrations and the plasma-generating gas. Just as the plasma transfers energy to other gas molecules by collision, it transfers energy to nearby surfaces including the substrate to be coated. Consequently there is some tendency for the substrate to become heated indirectly by the plasma and so it is necessary to limit the plasma powers used in low temperature CVD. Also the generation of a plasma in close proximity to the substrate can lead to the removal of material by sputtering. This can be exploited as a preliminary cleaning method prior to coating, but during coating the energies must be restricted to avoid sputtering the deposit away as fast as it forms. The increase in chamber pressure, which generally occurs on the introduction of the CVD reactants, tends to suppress loss of coating by sputtering.

Either D.C. or A.C. voltages can be used to generate a plasma with either capacitive or inductive coupling. The choice depends upon the type of electrode geometry most suitable for the coating application. Flat plates are easily coated in a parallel plate arrangement, whereas cylindrical objects can be handled inside a concentric electrode arrangement. In special cases a coil has been placed around the tube to be coated.

PACVD Equipment

The geometry of PACVD reactors depends upon the type of component to be coated and the method of plasma generation. For electronic substrates, which are usually in the form of thin discs, a parallel plate arrangement is best (Figure 2a). Either A.C. or D.C. power can be used. The pressure is usually maintained at a value 0.01-1 mB to obtain a stable plasma which is not swamped by the reactants. The gas inlet is arranged in the centre of the plates so that a radial gas flow is obtained over the surfaces to be coated.

In the special case of coating tubes a radio-frequency powered coil is placed around the tube (Figure 2b). By appropriate control of the gas pressure within the tube a localised plasma may be established inside the tube which can be made to transverse up and down the tube along with the coil[6].

Cylindrical objects may be coated inside a concentric tube arrangement (Figure 2c)[7]. They may form the central electrode in the case of large objects or be positioned around a permanent central electrode in the case of small objects.

Glass reactors are mostly used of PACVD so that there is direct observation of the plasma. The pressure region requires oil sealed rotary vacuum pumps, often with some kind of assistance (diffusion or blower).

Examples of Plasma Assisted CVD (PACVD)

Table 3 lists some PACVD reactions reported recently and the applications for which they are being developed. In all cases there is a reason associated with the application which has made it necessary to adopt a low-temperature route, although in some cases the temperature is still quite high.

PACVD of Oxides

Al_2O_3, SiO_2 and TiO_2 deposits have been prepared by passing the appropriate metal chloride and oxygen into a plasma.[6,8,9] The substrate temperatures used for Al_2O_3, and TiO_2 are 250-350°C. Al_2O_3 is an excellent passive encapsulant for electronic devices and TiO_2 is used as a dielectric in M O S devices. The deposition of doped SiO_2 layers is of great interest in the formation of optical fibres. In this case the plasma is an extremely convenient way of forming a deposit on the inside of a silica tube. By transversing an r.f. coil along the length of a silica tube it is possible to form glassy layers of doped SiO_2 on the inside surface without melting the whole tube (Figure 2c). The tube is subsequently collapsed and drawn out into a fibre of graded refractive index.

The use of oxygen in these reactions is made possible by the plasma. In the absence of a plasma, some kind of hydrolysis would have to be used to obtain the oxide coatings from the metal chlorides. This would result in residual OH groups in the deposits which would degrade the electrical and optical properties of the layers.

PACVD of Nitrides

The formation of boron and silicon nitride layers at low temperature under chemically

Table 3 PACVD process

Material Deposited	Chemical Reaction	Temperature of Substrate (^{o}C)	Application	Ref
Al_2O_3	$AlCl_3 + O_2$	250–350	Passive encapsulating layer for electronics	26
SiO_2	$SiCl_4 + O_2$	1000	Optical films	6
TiO_2	$TiCl_4 + CO_2/O_2$		Dielectric in M O S	22
BN	$B_2H_6 + NH_3$	400–700	Diffusion doping of Si	27
Si_3N_4	$SiH_4 + N_2/NH_3$	350	Encapsulation of 111 – V semiconductors	27
TiN	$TiCl_4 + N_2 + H_2$	400–600	Hard layer for wear resistance	28
SiC	$SiH_4 + C_2H_4/CH_4$		Electrochemical machining tools/ Thermal printer heads	29
TiC	$TiCl_4 + CH_4$	500–700	Hard layer for wear resistance	28
As	AsH_3		111 – V semiconductors	30
Mo	$Mo(CO)_5$		Antireflectance coating	31
Ni	$Ni(CO)_4$			32
Si	SiH_4	300	Semiconductors	33

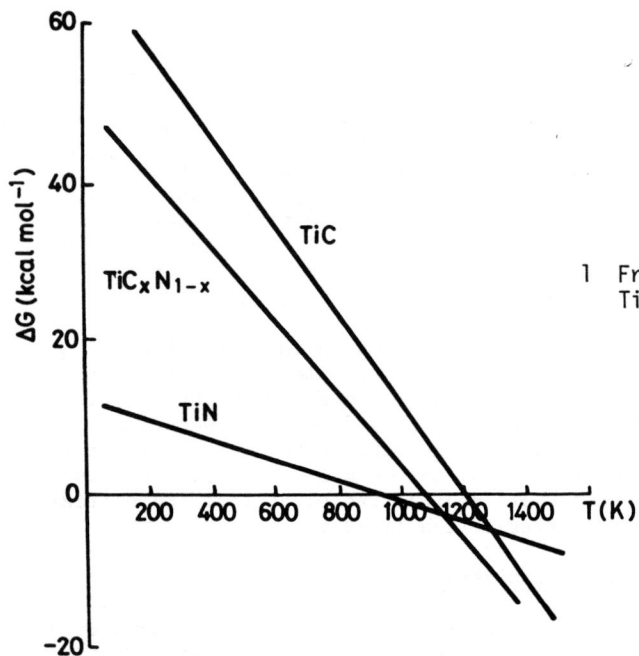

1 Free energy of formation of TiC, TiN and TiC_xN_{1-x}

'clean' conditions has electronic applications.[10] Layers of BN can be used in the solid state diffusion doping of silicon and silicon nitride is used as an encapsulating layer for 111-V semiconductor devices. The reaction between diborane or silane and ammonia in the presence of a plasma produces the layers required.

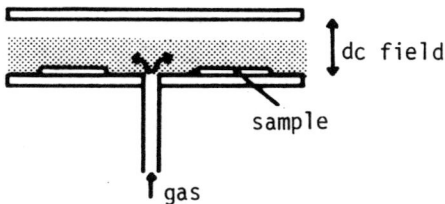

$$B_2H_6 + NH_3 \xrightarrow{\text{Plasma}} BN + H_2$$

$$SiH_4 + NH_3 \xrightarrow{\text{Plasma}} Si_3N_4 + H_2$$

There are no aggressive halide-containing by-products which might attack the electronic substrate. In the absence of the plasma these reactions would not occur until higher temperatures, and would tend to give mixtures of hydrogen-containing adducts rather than pure BN and Si_3N_4.

Titanium nitride can be obtained from the reaction between $TiCl_4$, N_2 and H_2 in the presence of a plasma at temperatures in the range $300-600^\circ$C, whereas in the absence of a plasma a temperature of $800-1000^\circ$C is required [11]. Titanium nitride layers have been investigated for their electronic properties and also for their potential as wear resistant layers. The substrate temperature has no effect on the rate of deposit, but does appear to influence the bond strength between coating and substrate.

PACVD of Carbides

Silicon carbide coatings are of interest when an oxidation and erosion resistant electrical conductor is required.[12] It has been used in the elements of heat activated printers and on electro-chemical machining tools. Silicon carbide can be obtained by conventional CVD, but the deposition temperature required is in excess of 1200°C, and this is impractical for many applications. However, in the presence of a plasma, silane and methane or ethylene will react to form a layer of amorphous silicon carbide.

$$SiH_4 + CH_4 \longrightarrow SiC + H_2$$

Again this is a very clean process which can be applied by any substrate without concern about the corrosive nature of the reactants or by-products.

PACVD of metals and elements

As yet PACVD has only been investigated for a few metals, but it has clearly got potential, particularly for the decomposition of carbonyls and hydrides. The deposition of molybdenum from $Mo(CO)_5$ has been investigated.[13] This is an interesting case as with conventional CVD a mixture of molybdenum and molybdenum carbide is invariably obtained.

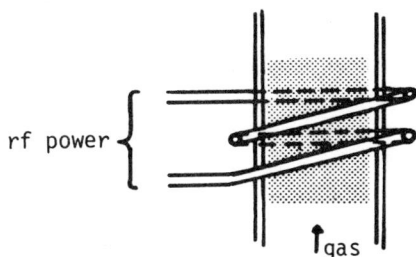

(a) Parallel plate electrodes

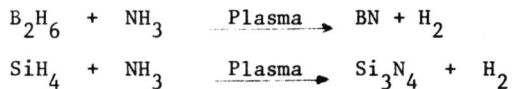

(b) Tube sample

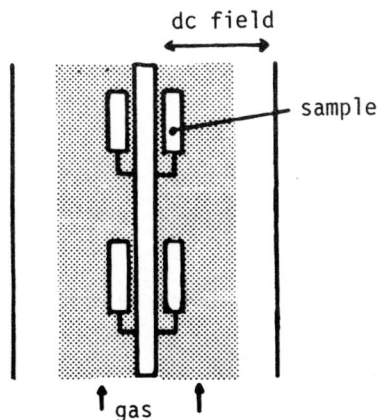

(c) Concentric electrodes

2 Electrode configurations for PACVD

Conclusion

PACVD is a technique which has already established a place in the manufacture of electronic devices. However, it has yet to be be ranked with respect to all of the PVD methods under investigation for the preparation of wear resistant surfaces. PACVD is technically simpler than many of the PVD methods, and has the high

volume throughput characteristics of CVD.
If it gives a product which is technically
equal to an application, then it is likely
to be one of the lower-cost options.

REFERENCES

1   Proc. 8th Int. Conf. on CVD,
    The Electrochemical Society, 1981.

2   Deutsche Edelstahlwerke Aktiengesellschaft
    U.K. Pat. 1 332 878, March 1972.

3   A. J. Perry and N. J. Archer, AGARD Lecture
    Series No. 106, NATO, Paris, 1979.

4   J. B. Mullin and S. J. C. Irvine,
    J. Vac. Sci. Technol., 1982, 21, 178.

5   H. F. Starling et al., Vide, 1966, 21, 80.

6   J. Irven and A. Robinson,
    Electronics Letters, 1979, 15, 252.

7   K. R. Linger, Proc. 9th Plansee Seminar,
    1978, Paper D12.

8   N. J. Archer, in "High Temperature Chemistry
    of Inorganic and Ceramic Materials", Chemical
    Society Special Publication No. 30, 1977.

9   M. J. Hakim, Proc. 5th Int. Conf. on CVD,
    The Electrochemical Society, 1975.

10 J. Gebhardt et al., Proc. 5th Int. Conf. on CVD,
    The Electrochemical Society, 1975, 786.

11 R. G. Bourdeau, U.S. Pat. 3 334 967, 1967.

12 H. Beutler et al., Proc. 5th Int. Conf. on CVD,
    The Electrochemical Society, 1975, 749.

13 C. Hollabaugh et al., Proc. 6th Int. Conf. on
    CVD, The Electrochemical Society, 1977, 419.

14 H. O. Pierson, Thin Solid Films, 1977, 45, 257.

15 K. Tamaru, J. Phys. Chem., 1955, 59, 777.

16 R. B. Reeves and J. J. Gebhardt, SAMPE, 10, D13.

17 D. J. Dumin et al., RCA Rev., 1970, 31, 620.

18 K. Hieber and M. Stolz, Proc. 5th Int. Conf.
    on CVD, The Electrochemical Society, 1975, 436.

19 Powell et al., p. 290, "Vapor Deposition",
    John Wiley, 1966.

20 J. Graham, High Temperatures-High Pressures,
    1974, 6, 577.

21 D. E. Clark et al., Proc. 8th Int. Conf. on
    CVD, The Electrochemical Society, 1981, 699.

22 C. C. Wang et al., RCA Rev., 1970, 31, 728.

23 R. Warren and M. Carlsson, Proc. 5th Int. Conf.
    on CVD, The Electrochemical Society, 1975, 611.

24 Powell et al., p. 362, "Vapor Deposition",
    John Wiley, 1966.

25 N. J. Archer and K. K. Yee, Wear, 1978, 48, 237.

26 H. Katto and Y. Kogo, J. Electrochem. Soc.,
    1980, 118, 1619.

27 K. Shohna et al., J. Electrochem. Soc.,
    1980, 127, 1546.

28 N. J. Archer, Thin Solid Films, 1981, 80, 221.

29 G. Verspui, Proc. 6th Int. Conf. on CVD,
    The Electrochemical Society, 1977, 366.

30 J. C. Knights and J. E. Matan,
    Solid State Corrosion, 1977, 21, 983.

31 B. O. Seraphin, J. Vac. Sci. Technol.,
    1979, 16, 193.

32 R. L. Van Hemert et al., J. Electrochem. Soc.,
    1965, 112, 1123.

33 H. F. Sterling and R. C. G. Swan,
    Solid State Electronics, 1965, 8, 653.

A P WEBB and D J FABIAN

Plasma transport and deposition of carbon films

SYNOPSIS

A technique for the deposition of thin-film materials is described. Plasma induced chemical transport provides an excellent method for performing high-temperature chemistry at low temperatures. A description of the experiments for deposition of carbon is given and the characterisation and some of the properties of the films using SEM, XPS and x-ray diffraction techniques are discussed.

THE AUTHORS

are in the Department of Metallurgy, University of Strathclyde, Glasgow, Scotland. Dr Webb is at present in the Department of Electronics and Electrical Engineering at the University of Glasgow.

INTRODUCTION

A number of experimental studies have been reported on the preparation of carbon films using glow discharge techniques. However, interest in the formation of carbon films is not limited to thin film technology; for example, an understanding of the interaction of carbon with hydrogen under plasma conditions, has been sought for many years for the purpose of using graphite and carbon materials as limiters in Tokamaks. In the same field, the use of a hydrogen discharge for cleaning Tokamak walls is currently in use, and since oxygen and carbon form the main contaminants studies of the basic reactions of hydrogen with oxygen and with carbon are important. Carbon and silicon plasma-induced deposits are also currently under investigation for the formation of wear-resistant coatings.

Glow discharge methods, to form carbon layers, divide roughly into two categories; composite gas decomposition or dissociation; (see for example Holland and Ojha[1] and references therein); and secondly, rf plasma induced chemical transport[2] (or PICT). The various methods are difficult to compare for carbon because the product materials formed are sometimes different and also because the plasma parameters employed are often confusingly reported.

However, we can make such a comparison in the case of silicon. Films of amorphous silicon (a-Si) have, in most cases, been formed on decomposition of silane in an rf-activated discharge[3], while microcrystalline silicon (μc-Si) has been produced by chemical transport in both rf and dc hydrogen discharges[4-5]. Deposition temperature is important because it can affect both the 'economics' of deposition and the properties of the deposit. For example, silicon has been deposited using PICT[6] at temperatures as low as 80°C. The properties of the materials also change on variation of other deposition parameters; control over these governs for example the lattice constant and crystallite size. Thus a transition state, amorphous to microcrystalline (a-μc), and the consequent changes in properties of deposited film, can readily be studied[7-8].

Turning to carbon, we report here an examination of films produced by PICT in a dc discharge, and characterisation of the material formed, for later comparison with the material prepared by decomposition in an rf-activated plasma.

GENERAL

A chemical transport reaction is one in which a condensed phase first reacts with a gas phase to form exclusively vapour phase reaction products and these in turn undergo the reverse reaction at a different location in the system, resulting in re-formation of the condensed phase. A simple form of such a reaction is:

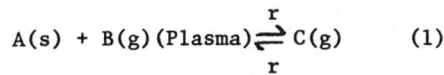

$$A(s) + B(g)(Plasma) \underset{\underline{r}}{\overset{r}{\rightleftharpoons}} C(g) \qquad (1)$$

in which solid A reacts with the gas discharge B to form gas C at one location, and then — because of a shift in the chemical equilibrium — the reverse reaction occurs at another location where condensed A is deposited. r and $\underline{r}$ are the rates of the forward and reverse reactions respectively. The change in the equilibrium conditions required, for the reversal of a heterogeneous reaction in this way, can be achieved by plasma parameter gradients, and by temperature or pressure differences. The transport of material may appear to be a process of sublimation or evaporation, but the solid substance transported does not have an appreciable vapour pressure at the applied temperature and it is chemically transported in the plasma.

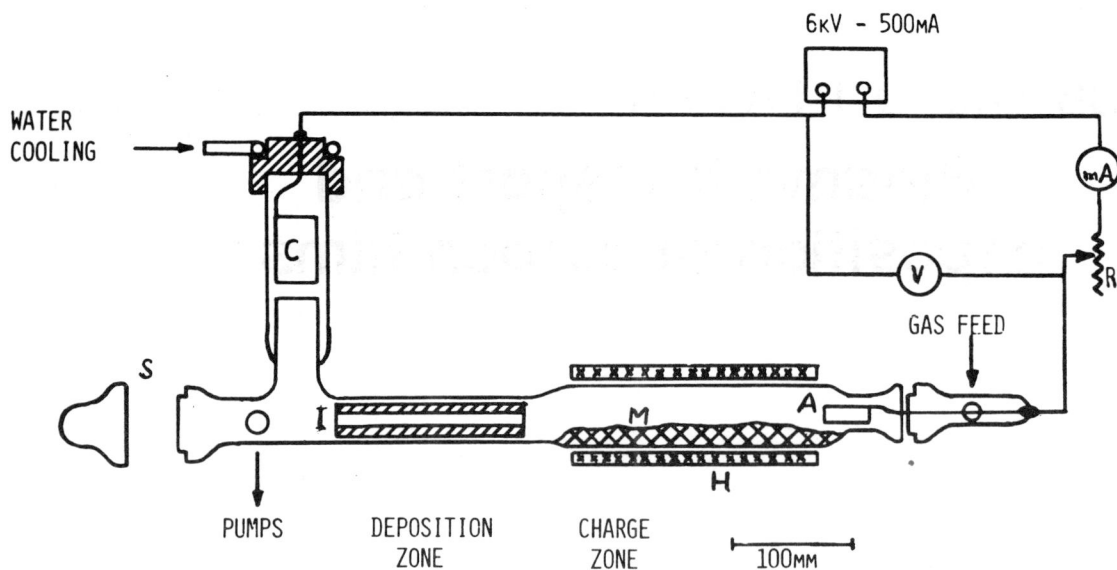

Fig. 1   Experimental arrangement for plasma
induced chemical transport of carbon

A/C – Molybdenum anode/cathode
M   – Charge material
H   – External heaters (or coolers)
I   – Quartz inserts
S   – Demountable vacuum seal

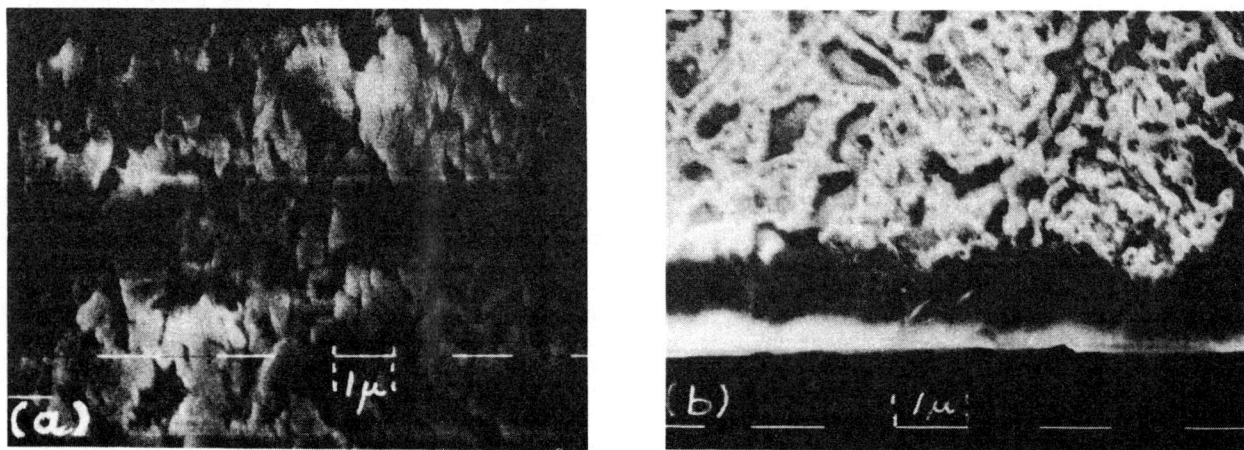

Fig. 2   Scanning electron micrographs of
nitrogen-plasma-deposited film:
(a) as deposited, (b) after Ar$^+$
etching, showing also edge of film.

In describing transport systems the following conventions are followed[9]:

1) the condensed substance to be transported is written on the left side of the chemical equation

2) if transport is accomplished by applying a temperature gradient , then T_2 always denotes the higher and T_1 the lower temperature; i.e. $T_2 > T_1$

3) an arrow is used to denote the direction of transport, e.g. $T_1 \rightarrow T_2$ denotes transport from a lower temperature to a higher temperature zone.

The two specific systems studied, carbon transport by hydrogen and by nitrogen plasmas, demonstrate clearly that transport can be achieved in both exothermic and endothermic plasma reactions. The exothermic system

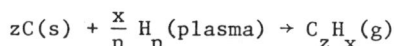

$$zC(s) + \frac{x}{n} H_n \text{(plasma)} \rightarrow C_z H_x \text{(g)}$$

proceeds in a low-energy zone (E_1), and the reverse decomposition occurs in a high-energy zone (E_2); i.e. with $E_2 > E_1$, carbon is deposited in the E_2 zone. The transport has been attributed to the formation and subsequent decomposition of simple hydrocarbon radicals[2,10,11].

The reaction of carbon with nitrogen on the other hand is strongly endothermic, and carbon is transported in nitrogen in the direction of decreasing plasma energy $E_2 \rightarrow E_1$.

$$C(s) + \frac{1}{x} N_x \text{(plasma)} \rightarrow CN \text{(g)}$$

The carbon can thus be deposited in a low-energy zone.

EXPERIMENTAL

The quartz tube (Fig. 1), 50mm diameter, pumped by oil (santovac) diffusion and rotary pumps with a liquid nitrogen (LN_2) trap, is evacuated to a base pressure of 10^{-5} torr, monitored by pirani and penning gauges. Oxygen-free hydrogen (purity 5 9's) is admitted to the system via a LN_2 trap through a fine-control needle valve. Nitrogen and argon also can be admitted through the needle valve, as alternative on-line gases. The gas pressure is monitored with an oil (santovac) "U" manometer.

The gas enters the reaction tube in the charge zone. The charge material M consists of UCAR carbon fines (supplied by Union Carbide) of particle size 80 to 420μm, or carbon rods 6mm diameter and length 150mm. The deposition zone, downstream, is of a constricted diameter to create a more intense discharge with respect to the charge zone, depending on the amount of carbon in this zone.

Quartz inserts, I, or substrates can also be inserted through the seal at S to accommodate deposited material, and substrates with deposits are easily removed for characterization and analysis. The inserts are used also to create different current densities for deposition by restricting the deposition zone. External cool-air blowers, or alternatively tape heaters H, are used to regulate the temperature in each zone.

The plasma is sustained between two cylindrical molybdenum electrodes A and C. The cathode, mounted directly on a metallic flange, is situated downstream, close to the pumping exit of the tube so as to remove any impurities or contaminants introduced by cathodic processes. It is also water-cooled to promote a stable discharge, which is sustained by a 6kV 500mA dc power supply and high-power stabilizing resistor chain R.

RESULTS

1. Carbon-hydrogen system

The reactor tube was operated such that transport from zone E_1 to E_2 occurred, where $E_2 > E_1$. Both forms of carbon were tried, the powder form giving the better deposits.

The high-energy deposition zone was formed using quartz inserts concentrically placed to constrict the discharge and so increase the current density. Soda-lime glass and molybdenum substrates were used within the inserts to collect deposited material for characterization. Annealing of the charge region under vacuum prior to exciting the discharge was used to clean the carbon.

Experimental conditions during deposition runs were in the following ranges: pressure 0.8 to 1.8 torr, flow rate 10 to 70 torr ml sec^{-1}, and discharge current 80 to 360 mA, which imply current densities of 1-2 amp cm^{-2}. Temperature in the zone was not measured directly but was monitored from outside the reaction tube wall. Deposition was tried using external heating and also by relying on high current density to provide the high-energy for decomposition. External heating is only effective when working at higher pressures, and also when the concentric inserts fit closely inside one another so that constriction of the discharge in the deposition zone is effective; leakage around the sides or between inserts serves to lower the high current density.

No characteristic carbon layers were deposited on substrates investigated to-date, but layers ranging in colour from pale yellow to black were deposited in inserts of diameter down to $\sim$3mm.

2. Carbon-nitrogen system

For carbon transport in a nitrogen plasma, the reactor tube conditions were changed to cause transport from E_2 to E_1, and quantities of material in the charge region were increased to give a more intense discharge; no inserts were used in the deposition zone, but plane substrates ($\sim$5mm x 10mm) were used to collect the deposited material in thin film form. Additional heating was also supplied to the charge region, raising the outside wall to over 600°C, at pressures >1 torr, with currents of 200mA and above.

Preliminary results with this system show that carbon can be transported as predicted, and can be made to deposit on the walls of the tube as well as on glass and plastic (PVC) substrates. Thin film deposits were obtained

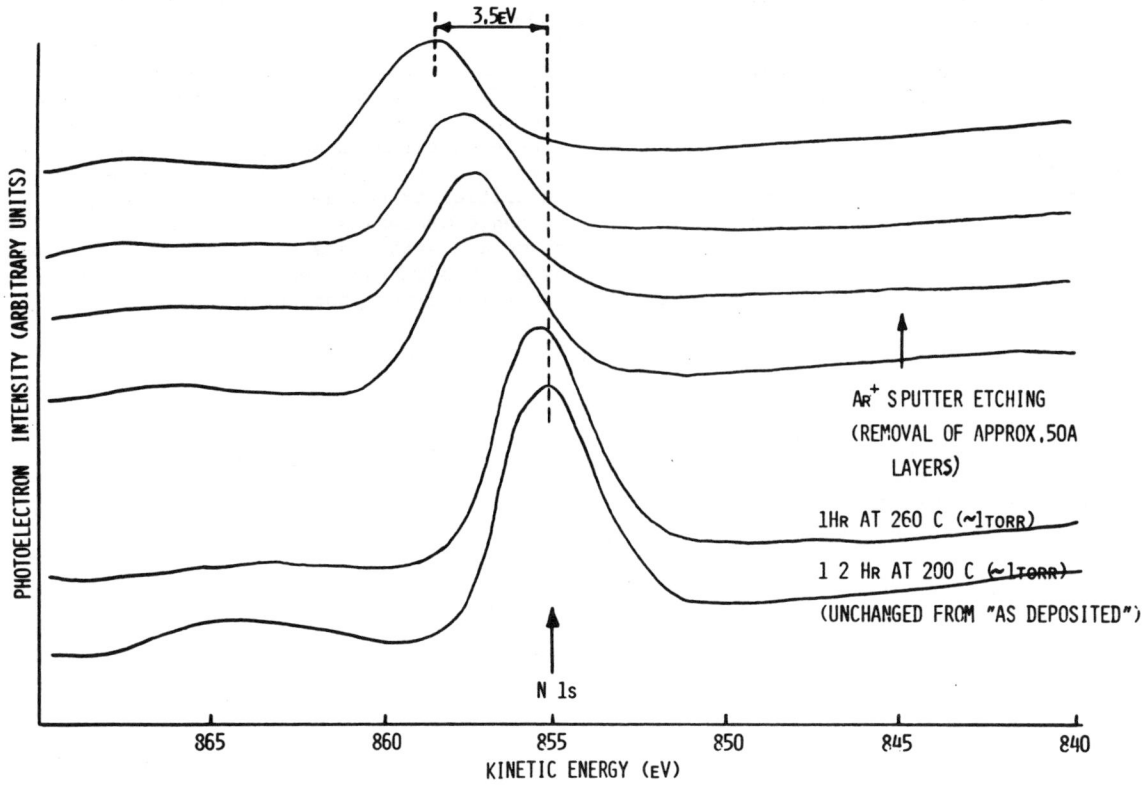

Fig. 3   X-ray photoelectron spectra showing
the shift of the N 1s peak after
recorded periods of Ar⁺ etching for
a ∿0.5μm film of carbon deposited
on a glass substrate.

Fig. 4   3-stage differentially pumped
modulated-beam quadrupole mass
spectrometer system for measurement
of transient species in plasma
reactor.

with pressures in the reactor from 0.4 to 1.7 torr, and currents in the range 80 to 240mA. Flow rates were in the same range as quoted for the hydrogen plasma. Substrate temperature measurements were made at different positions in the deposition zone at the relevant pressures and currents, and ranged from 45 to 130°C. The flux of N atoms in the deposition zone was calculated as $\sim 1 \times 10^{19}$ $cm^{-2}sec^{-1}$ for most films deposited, assuming dissociation probabilities of the order of only a few per cent from molecular nitrogen.

DISCUSSION

Carbon-hydrogen deposition

Ojha et al[12] have reported the growth of carbon films by dissociation of various hydrocarbon gases; they examined the effect of hydrogen-to-carbon ratio on the rate of growth and the properties of the deposit. Carbon transport in an intense hydrogen discharge has also been studied, a high current density being used to decompose the hydrocarbon radicals[2]. Both these investigations employed an rf-excited discharge.

In the present work (using a dc hydrogen discharge) layers deposited in the inserts have not to-date been characterized because tubes of diameter $\sim 3mm$ had to be used to create the high current density necessary to form the carbon (black) films. The pale yellow deposits formed in some instances are polymers of high hydrogen content and result from the dissociative energy being too low to decompose completely the C_2H_x molecular and radical species in the plasma. It has not been possible to characterize any of the deposits to-date, but their existence demonstrates that plasma assisted chemical transport can be achieved in hydrogen.

On the other hand the experiments in which no deposition was observed show that the temperatures required to achieve the necessary high energy for carbon deposition are above the devitrification point of the substrate glass used, $\sim 815°C$ in this case. Metal surfaces such as molybdenum have a high efficiency for catalytic recombination of atoms and meta-stable species, leading to a decrease in plasma energy near to the deposition surface[13]. The deposition conditions for this system are being further explored.

Carbon-nitrogen deposition

Thin film deposits were successfully grown from a nitrogen discharge on glass and plastic substrates, in the pressure range 0.4-1.7 torr, with currents from 80 to 240mA. Optimization of the flow rate was found to be of paramount importance in agreement with previous reports[14].

The deposits obtained adhere well to both glass and plastic (polypropylene) surfaces; they appear both wear-resistant and chemically inert. Characterization of the deposits, which were predominantly amorphous, was achieved by x-ray diffraction (X-D), scanning electron microscopy (SEM) - see Fig. 2 - and x-ray photoelectron spectroscopy (XPS).

The XPS investigation was initially undertaken to examine the bonding at the film-to-substrate interface. C, N and O 1s peaks were monitored after successive layers had been removed from the surface of the film by Ar^+ ion sputter etching. The thickness of each layer removed could be calculated from Ar^+ sputtering data for glassy carbon[15] (i.e. amorphous carbon), adjusted to take account of the angle of incidence. Ar^+ 'sectioning' in this way resulted in a broadening and a shift of the peaks to lower binding energies, following each treatment. The shift from the accepted calibration peak energy was as much as 3-4eV, and to higher kinetic energies; whereas a monitored Ag 3p peak (from silver conducting resin-cement used for bonding substrates to the sample holder) remained unchanged. The argon ions, at the energy used, result not only in an appreciable sputtering yield, but can also penetrate into the film. An EDAX spectrum clearly shows the presence of the argon, whereas XPS Ar peaks were not detectable (as has been the case in other investigations[16]). The effect is under detailed investigation. Preliminary results appear to indicate the possibility of some 'chemical' bonding between film and substrate (Fig. 3) and also to imply formation of a dipole layer in the film on Ar^+ etching due to penetration and lodging of Ar^+ ions in vacancies.

Films on the plastic substrates which are very similar in appearance to those grown on glass were grown at low discharge currents 120-80mA, which keeps the deposition temperature to a minimum, and subsequently the substrates show no sign of deformation. Negative weight changes after deposition occurred and probably indicate some kind of preliminary leaching action in the plasma before the dominant active transport and deposition species takes over and deposits the film. An XPS study is now also in progress on these films.

FURTHER STUDY

The conditions of film deposition are clearly important in explaining why material is produced in different forms with different properties and it is therefore advantageous to study in depth the fundamental chemical reactions of the system used for deposition.

For a reaction of the form (1), chemical equilibrium is achieved when the rates of the forward and reverse reactions are equal, ($r = \underline{r}$). In general $r > \underline{r}$ leads to pick-up of charge in the reaction zone and $r < \underline{r}$ to deposition. The term partial chemical equilibrium (PCE) is used to describe chemical equilibrium under non-isothermal plasma conditions. Starting from a non-equilibrium state, PCE is approached at a rate characterized by a time τ such that

$$C(g)\big|_{t=\tau} = \frac{1}{e}\left\{ C(g)\big|_{t=0} - C(g)\big|_{t \to \infty} \right\}$$

where t is the residence time of the species in the reaction zone. The departure of the system from PCE depends on the ratio t/τ. For example, $t/\tau \gg 1$ implies that the system is close to PCE. The value of τ is a measure of the overall reaction rate, i.e. a high reaction rate corresponds to a small τ.

In the case of the silicon-hydrogen system, it has been shown that the dissociation of the transient species, $C(g)$, using PICT, to form μc-Si, occurs close to the equilibrium, whilst deposition of a-Si from silane takes place far away from chemical equilibrium[17].

Knowledge of the kinetics of the system has been obtained using thermogravimetry[18], mass spectrometry and a plasma sampling system[19], and gives information concerning the degree to which the system departs from chemical equilibrium.

In our laboratories, a three stage differentially pumped plasma sampling reactor (constructed by Leisk Engineering Co.) - see Fig. 4 - coupled to a ENL quadrupole mass spectrometer (Spectrum Scientific Co.), has now been assembled and is to be used for an examination of the transient species in various plasma-solid systems. This information coupled with lifetimes of the species and rates of transport should provide further insight on the transport reaction, and be of importance in the control of thin-film deposition.

ACKNOWLEDGMENTS

We should like to thank Dr. S. Veprek (University of Zürich) for his comments and interest in this work. The research was supported by the University of Strathclyde Research and Development Fund, and by the Science and Engineering Research Council.

REFERENCES

1. L. Holland and S.M. Ojha, Thin Sol. Films 58 (1979) 107.

2. S. Veprek, Journ. Crystal Growth 17 (1972) 101; and Z. Phys. Chem. N.F. 86 (1973) 95.

3. H.F. Sterling and R.C.G. Swann, Sol.St. Elect. 8 (1965) 653.

4. S. Veprek and V. Marecek, Sol. St. Elect. 11 (1968) 683.

5. Z. Iqbal, A.P. Webb and S. Veprek, Appl. Phys. Lett.36 (1980) 163; and S. Veprek, Chimia 34 (1980) 489.

6. S. Veprek, Z. Iqbal, H.R. Oswald and A.P. Webb, J. Phys. C : Sol. State Phys. 14 (1981) 295.

7. S. Veprek, Z. Iqbal, H.R. Oswald, F.A. Sarott, J.J. Wagner and A.P. Webb, Solid State Commun. 39 (1981) 509.

8. S. Veprek, Z. Iqbal and F.A. Sarott, Phil. Mag. 45B (1982) 137.

9. H. Schafer, Chemical Transport Reactions, (Academic Press 1964).

10. S. Veprek and W. Peier, Chem. Phys. 2 (1973) 478.

11. S. Veprek, D.L. Cocke and K.A. Gingerich, Chem. Phys. 7 (1975) 294.

12. S.M. Ojha, H. Norstrom and D. McCulloch, Thin Sol. Films 60 (1979) 213.

13. S. Veprek, Topics in Current Chemistry 56 (1975) 139.

14. K.G. Muller, Chem. Ing. Tech. 45 (1973) 122.

15. B.M.U. Schertzer, R. Behrisch and J. Roth. Proc. Int. Symp. Plasma Wall Interactions, Julich FRG (1976) p.353.

16. G.A. Sawatzky reports (private communication) that it is difficult to detect argon, even at very low sputtering energies where any implanted argon should be very close to the film surface.

17. S. Veprek, Z. Iqbal, H.R. Oswald, F.A. Sarott and J.J. Wagner, J. Phys. (Paris) 42 (1981) C4-251.

18. A.P. Webb and S. Veprek, Chem. Phys. Lett. 62 (1979) 173.

19. J.J. Wagner and S. Veprek, Plasma Chem. and Plasma Proc. (1982) In press.

H I PHILIP and D J SCHIFFRIN

Glow discharge assisted chemical vapour deposition

SYNOPSIS

The deposition of titanium and chromium from volatile halides has been carried out using a glow discharge. The process is equivalent to the CVD technique, but the substrate temperature can be kept below 500°C since the gas phase reactions are driven by energy transfer from the electrons in the cathodic region to the reacting gas. The advantages of this coating technique are: low substrate temperatures, high throwing power, wide range of coatings and the possibility of having a range of process control variables such as pressure, potential and composition. The detailed reaction mechanisms have not been elucidated, but evidence is presented in favour of a purely chemical process; it is also shown that a simple thermodynamic analysis can give useful information regarding the feasibility of carrying out reactions leading to specific coatings in the glow discharge.

THE AUTHORS

are in the Chemistry Department at Southampton University, England.

INTRODUCTION

Chemical vapour deposition (CVD) is a very versatile and powerful coating technique, which has been extensively employed for the production of wear resistant, corrosion resistant or refractory coatings.[1-5] One of the main drawbacks of CVD is the high temperatures that are often required to drive thermally the desired reactions.

Archer[6] proposed the use of plasma excitation to avoid this problem, and reported efficient plasma-assisted CVD of TiC, TiN and TiC_xN_{1-x} when the substrate temperature was kept in the range 400-600°C. The idea was to make use of the high energy inherent in either a radio-frequency or a D.C. gas discharge to produce the desired chemical reaction in the plasma region, the products of which subsequently could be deposited on the cool substrate.

The technique of r.f. plasma deposition has been extensively studied previously, for instance for Si and SiO_2 deposition,[7,8] for the growth of silicon nitride films,[9] the formation of ultra-thin polymer coatings on electrodes,[10] etc. R.f. discharges generate homogeneously the reactive species or reaction products, which must diffuse to the surface to be coated. From this point of view the glow discharge presents, in principle, some advantages with respect to the homogeneous discharge methods since the reaction can be confined to the cathode space.

The purpose of the present work was to attempt to exploit the unique advantages offered by a glow discharge to perform chemical reactions in a well defined region of space very close to the work piece. It was hoped that this would result in the controlled deposition of metals, while keeping the substrate temperature below 500°C.

GENERAL CHARACTERISTICS OF GLOW DISCHARGE

The term glow discharge is usually employed to describe electrical conduction phenomena through a gas in the pressure range 0.1 to ~50 mbar. During the passage of current, several regions of interest are observed:[11] (1) the cathodic dark space or Crookes dark space; (2) a luminous region called the negative glow; (3) a dark space called the Faraday dark space; and (4) an anode glow region which is bound by a narrow anode dark space close to the positive electrode. The regions of interest to this work are (1) and (2), where the interaction between the energetic particles formed in the discharge and the gas molecules should result in the desired chemical reactions in a localised region very close to the cathode.

In the steady state, secondary electrons are emitted from the cathode by collision of gaseous cations with the surface; also photoemission from the cathode by the radiation generated in the discharge can occur. The electrons leaving the cathode are accelerated by the field and transfer energy to the gas by either elastic or inelastic collisions, some of which result in ionization of the gas.[12] The first part of this process occurs in the cathodic dark space and gas ionization results in an increase in the number of electrons and therefore in an enhancement of the rate of energy transfer. This multiplication mechanism is responsible for maintaining a stable discharge. Most of the potential drop across the discharge occurs in the cathodic region and the electric field decreases linearly with distance across the Crookes dark space.[13] The details of the potential distribution are outside the interest of this work; however, what does matter from the point of view of any gas phase chemical process occurring in the glow discharge is the rate of energy transfer to the gas and the spatial distribution of species formed.

CHEMICAL REACTIONS IN GLOW DISCHARGE

Two types of plasma have been considered as chemical reacting media: the thermal (also called high pressure, high power, arc or H field plasmas), and the cold plasmas.[14] The difference between these two electrical discharges is in the rate of

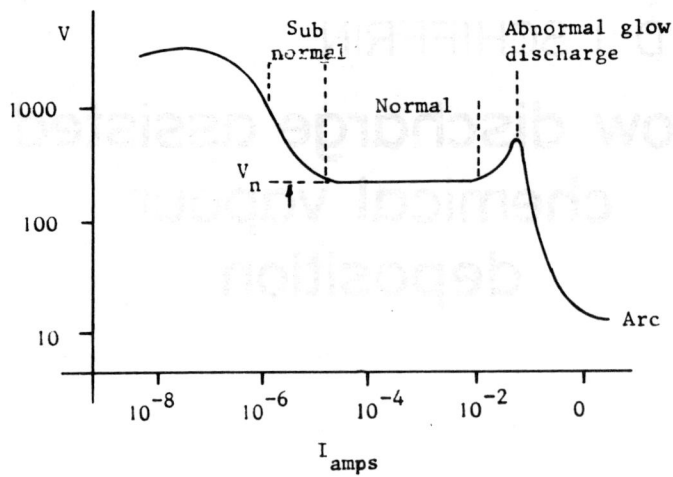

1 General current-voltage characteristics of
   discharge through gas at low pressures

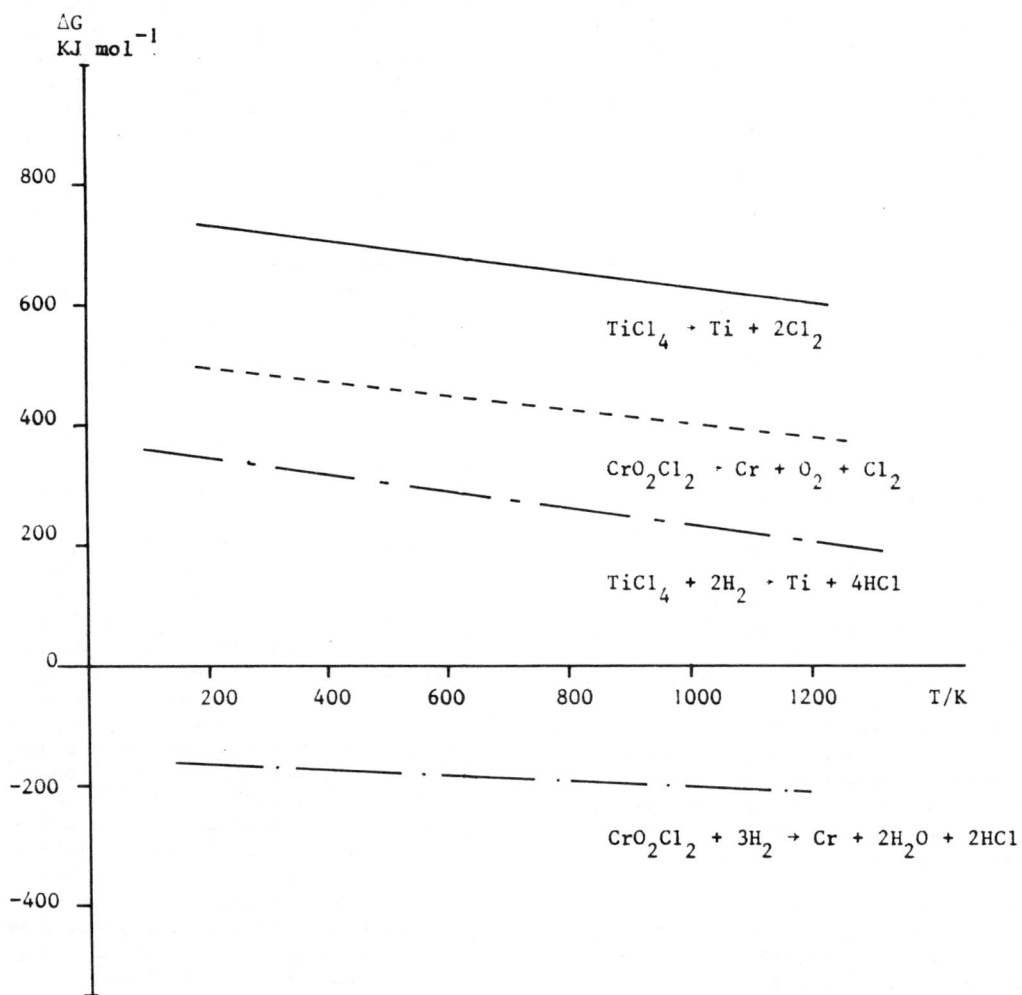

2 Free energy-temperature relationships for several possible gas phase
   chemical reactions leading to Ti and Cr metal deposition

energy transfer between particles and hence the degree of thermal equilibrium. In the high pressure thermal plasmas, thermal equilibrium deviations of only 0.01% have been measured and gas temperatures of several thousand degrees Kelvin can be obtained.[15] In contrast, in the "cold" plasmas generated, for instance, in a glow discharge the average temperature of the electron gas will differ significantly from that of the gas molecules due to the low efficiency of kinetic energy transfer between electrons and heavy gas molecules.[12] The electron gas energy can be calculated in principle for the rare gases and for simple molecules, where single step ionization is observed.[11] However, it is important to note that average electron energy values of 10 eV can be easily realized in cold discharges. Since the electron gas energy follows a Maxwellian energy distribution,[11] a wide range of molecules can be ionized by the discharge.

This feature of the glow discharge has been extensively used in synthetic chemical reactions, when the reaction requires highly reactive intermediates.[16,17] For instance Wartik et al[18] synthesized diboron tetrachloride from BCl_3; a variety of organic silanes have been prepared,[17] halogenation reactions have been performed,[16] and the surface polymerization of styrene has been carried out.[19] The formation of free radicals in sufficiently high concentration can result in second order radical recombination reactions, and this effect has been used by Boelhouwer and Waterman[20] to synthesize diphenyl, dibenzyl and other aromatic double ring systems from benzene, toluene, etc.

In the case of microwave discharges in gaseous mixtures, uniform reaction conditions can be achieved in the gas phase, and Cooper et al[21] have shown that the ratio E/P (E = electric field, P = gas pressure) is the parameter determining reaction conditions. These considerations give a clear indication that the glow discharge should be a powerful method for obtaining surface coatings by gas phase reactions. For the glow discharge in the cathodic region, besides the problem of evaluating energy transfer, an additional complication in the prediction of chemical effects is the spatial distribution of electrons having different energies. This is due to the linear decrease of electric field with distance observed.[13] These complications preclude at this stage a detailed analysis of the actual reaction environment. However, since the average electron energy will be linearly dependent on the cathodic field and hence approximately on the total potential applied across the glow discharge cell, it might be expected that the average reaction temperature in the cathodic region would be proportional to the applied voltage. This behaviour could give then a very simple control variable to the process.

Another control variable is the total gas pressure, either due to the reacting gas when this is present alone, or to the mixture of reacting and carrier gas employed. An important feature of the glow discharge is that the thickness of the cathodic fall region is inversely proportional to the pressure, i.e.[11,12]

$$p \, d_m = const.$$

where p = gas pressure and d_m = thickness of the cathodic fall region. Therefore it is possible to have some degree of control of the distance from the surface where the actual energetic chemical reactions occur. In most of the experimental work reported here to achieve metal deposition, the maximum pressure compatible with the absence of arcing was employed in order to ensure the highest possible rate of deposition.

Obviously, due to the voltage-current characteristics of the glow discharge (Figure 1) these variables cannot be altered independently of of each other in the <u>normal</u> glow discharge region,[11] and for this reason the <u>abnormal</u> glow discharge region was chosen for this work. In this current density range the cathodic dark space contracts as the current and potential are increased, affording thus a control on the operating conditions.

CHOICE OF GASEOUS COMPOUNDS

The deposition of titanium and chromium was chosen to study the general characteristics of the glow discharge processes; $TiCl_4$ and CrO_2Cl_2 were employed. Two different types of reaction were attempted: direct thermal decomposition and reduction with hydrogen.

The temperature dependence of the free energy of the relevant reactions is illustrated in Figure 2. The values of free energy at different temperatures have been calculated to a first approximation assuming that the heat capacitance terms in the variations of enthalpy and entropy are small, which is justified in this simplified treatment.[22,23] The appropriate thermodynamic data were taken from the literature.[24-26] From these data, the thermal decomposition temperatures calculated for $TiCl_4$ and CrO_2Cl_2 to give the metallic phases:

$$TiCl_4(g) \rightarrow Ti(s) + 2Cl_2(g) \qquad (1)$$

and

$$CrO_2Cl_2(g) \rightarrow Cr(s) + O_2(s) + Cl_2(g) \qquad (2)$$

were 6300 and 4250 K respectively. The corresponding temperature for the reduction of $TiCl_4$ with hydrogen:

$$TiCl_4(g) + 2H_2(g) \rightarrow Ti(s) + 4HCl(g) \qquad (3)$$

was 2450 K; the reduction of CrO_2Cl_2 is thermodynamically favourable at room temperature. For the reaction

$$CrO_2Cl_2 + 6H_2 \rightarrow Cr + 2H_2O + 2HCl \qquad (4)$$

$\Delta G^o_{298} = -167.4$ kJ/mole and $\Delta S^o_{298} = 53.1$ J/K mol. Therefore reaction (4) becomes more favourable as the temperature is increased. The four reactions studied covered the extremes of thermodynamic characteristics that can be encountered in practice.

Experimental

The experiments were performed in a conventional vacuum line. For the experiments without gas flow through the reaction chamber, this had a ballast vessel which could be filled with the selected compound or its mixture with a carrier gas (Ar or H_2); the reacting gas could then be admitted to the reaction vessel by means of a PTFE needle valve. The reacting gas partial pressure in the ballast was regulated by controlling the temperature of the $TiCl_4$ or CrO_2Cl_2 reservoir to $\pm$ 0.1°C using a circulating water thermostat.

The system could be evacuated using a rotary vacuum pump which was capable of reducing the pressure to 0.5 mbar in about 5 minutes. The system was purged with argon before the start of each run in order to ensure the exclusion of air from the system. The taps were of the conventional glass vacuum type.

The pressure in the system was measured roughly using a mercury manometer and more accurately (to about 0.2 mbar) in the range 1 to 10 mbar using a "Vacustat 1" McLeod gauge.

The electrical leads into the reaction vessel were sealed using epoxy resin which provided an

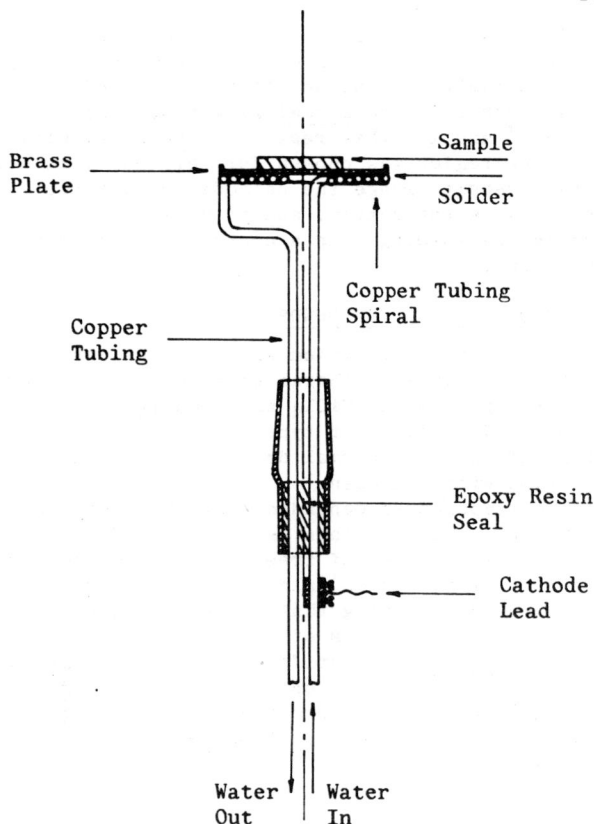

3 Diagram of sample holder used with flat coupons, incorporating sample cooling stage

Brass Plate
Sample
Solder
Copper Tubing Spiral
Copper Tubing
Epoxy Resin Seal
Cathode Lead
Water Out
Water In

effective vacuum seal, more durable than a metal-in-a-glass one. This method employed single strands of bare wire encapsulated in the epoxy as the vacuum seal; flexible connecting leads were then soldered onto the ends of the wire.

In the early experiments, with no cooling of the samples, it was found that the best results were obtained using a gas pressure of around 5 mbar. At higher pressure (> 8 mbar) the glow discharge tended to degenerate into an arc discharge which would rapidly burn the surface of the cathode. The pressure was found to change during the course of each experiment and therefore it was measured every few minutes and adjusted when necessary.

Power was supplied to the system using a Hartley Instruments Model 421 power supply. The output voltage could be varied between 0 and 1000 volts, and the output current could be controlled independently between 0 and 200 mA. An external digital voltmeter was employed in order to monitor the current.

During the course of this work, different coupons and reaction vessel geometries were used. Flat samples were employed in some experiments with $TiCl_4$; these cathodes were made of 0.25 cm thick flat steel cut into 2 x 2 cm squares. These were held on a sample stage which incorporated water cooling by copper tubing (Figure 3). The anode was made of titanium foil which was rolled into a cylinder about 0.2 cm in diameter and mounted in a glass tube. The electrical connection and seal was made with epoxy resin. The end of the anode was 0.5-1.0 cm from the cathode surface, but the distance between the anode and

cathode had no visible effect on the glow discharge, as expected for the pressure ranges studied.[12]

Cooling facilitated the use of higher currents (of 50 to 200 mA) while at the same time it minimised sputtering effects. When using the water cooled sample stage, it was found that much higher gas pressures (up to 35 mbar) could be employed without the discharge degenerating into an arc mode, and this effect was attributed to the improvement in sample cooling. The higher gas pressure had the advantage that more of the carrier gas could be accommodated within the system, and the volume of the vacuum line was increased by connecting a 1 litre flask to the reaction vessel. Increasing the pressure and the volume of the $TiCl_4$ vapour within the system helped to ensure that it would not be excessively depleted of titanium during the deposition process.

The disadvantage of the flat geometry was the difficulty in keeping a constant $TiCl_4$ concentration in the gas mixture. For this reason a cylindrical geometry was employed, with a system for keeping a flow of reacting mixture through the glow discharge. The water cooled reaction vessel and cathode holder is shown in Figure 4. This consisted of a tubular stainless steel liner 1.27 cm inside diameter and 2.6 cm long. A copper tube was wound round the outside of the liner through through which cooling water was passed during the discharge. The copper tube was soldered onto the steel in order to ensure good thermal contact. The length of the heat path between the inside of the steel tube and the cooling water was between 0.3 and 0.4 cm. The cathode holder was mounted in a ground glass B34 Quickfit cone using epoxy resin, and electrical contact was made to the holder through the copper tubing. The cone could be fitted in a Quickfit socket, with the anode mounted along its axis so that upon assembly the anode was positioned along the axis of the cathode. The anode consisted of a titanium foil cylinder 0.3 cm in diameter and 2.5 cm long held in the end of a pyrex glass tube. Upon assembly the concentric anode and cathode had an inter-electrode spacing of about 0.5 cm. A glass inlet tube was mounted at the lower end of the cathode holder and was provided with a ground glass joint so that a stopcock and metal halide reservoir could be connected to the vacuum line. The stopcock was of the PTFE screw type, thus allowing some regulation of the rate of gas flowing through the cathode bore. The cathode samples employed in these experiments consisted of nickel foil sheets 1.5 mm thick, 2.5 cm wide and 4 cm long. These were rolled into cylinders with a diameter about 1 mm greater than the inside diameter of the cathode holder, so that upon insertion into the holder their "springiness" held them firmly in contact with the inside of the holder, so ensuring good thermal contact.

In the initial runs the flow of the vapour through the cathode and the vapour pressure in the vacuum line were controlled only by means of the PTFE stopcock. The system was periodically re-evacuated with the vacuum pump, but this method proved unsatisfactory as the pressure inside the vacuum system could not be easily controlled and the system was subject to arcing due to changes in pressure which resulted in burning of the cathode surface.

The vacuum system was then modified using a thermostatted metal halide reservoir and cold trap. The temperature of the water jacket surrounding the halide reservoir was maintained at 9°C in the case of $TiCl_4$ and at 3°C in the case of CrO_2Cl_2 to give a vapour pressure of about 8 mbar. After its passage through the cathode the excess vapour was then condensed in the cold trap which was kept at

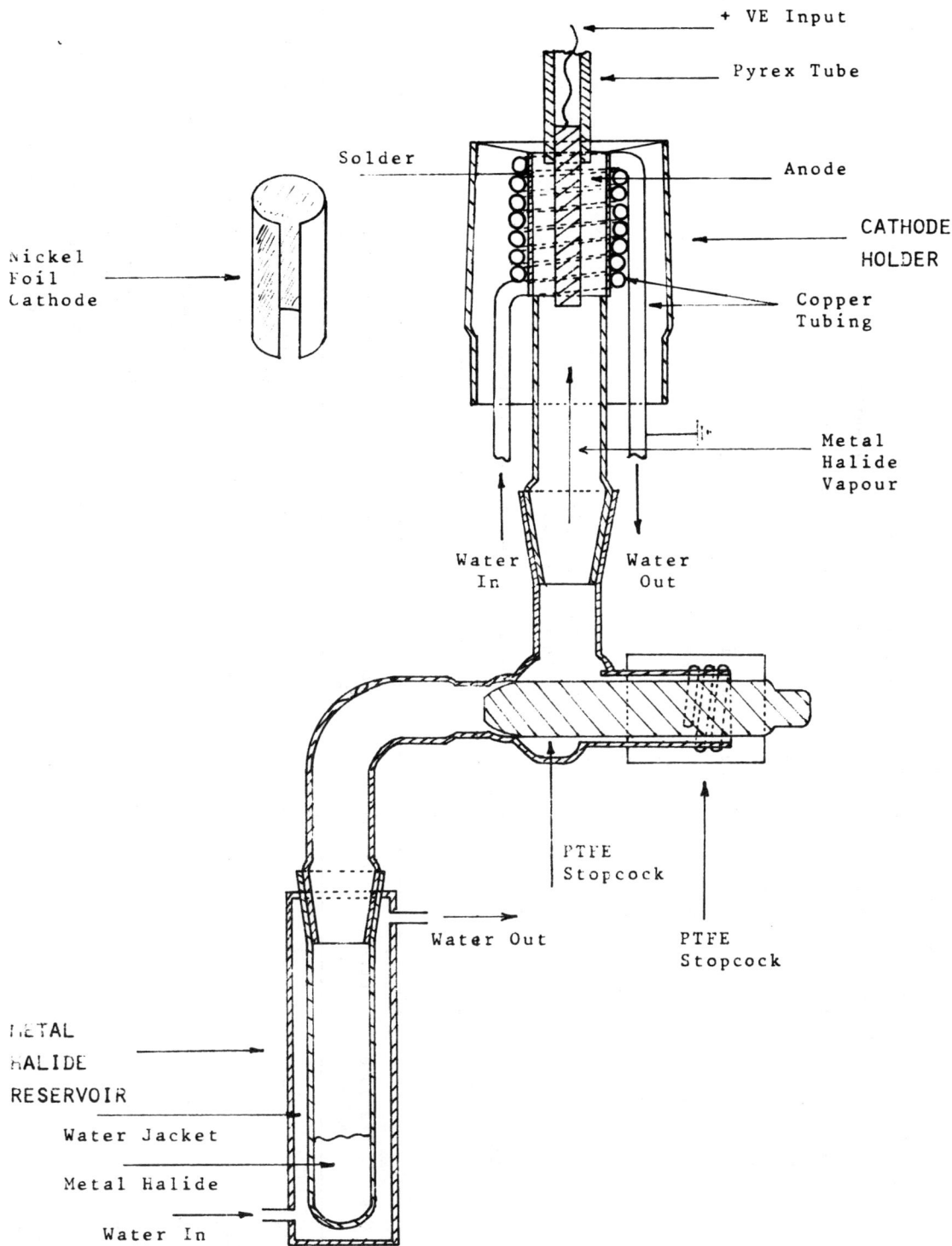

4 Diagram of cylindrical sample holder and volatile halide reservoir used for glow discharge experiments with flow of vapour through reaction cell

TABLE 1

TITANIUM DEPOSITION ON FLAT COUPONS

Sample No.	Gas	Pressure m bar	Current mA	Potential V	Time min	Weight gain mg	Sample Cooling	Microprobe Analysis Results
1	$TiCl_4$	3-5	5	500	60	0.3	None	0.1μm Ti dispersed over surface, peaks up to 1μm
2	$TiCl_4$	4	10	500	60	1.2	None	0.1μm Ti dispersed over surface
3	$TiCl_4$	5	20	550	60	2.7	None	0.2μm Ti dispersed over surface
4	$TiCl_4$	5	40	600	60	-7.1	None	
5	$TiCl_4$	5	40	600	60	-9.3	None	
6	$TiCl_4$	5	40	600	60	-5.2	None	
7	Argon	5-10	40	300	60	-0.3	None	No Ti detected
8	Argon	10	60	350	60	-4.7	None	
9	Argon	5	60	350	60	0	Water	
10	$TiCl_4$	4	40	600	60	-0.1	Water	
11	$TiCl_4$	5	60	500	60	1.0	Water	0.2μm Ti dispersed over surface, peaks up to 0.5μm
12	$TiCl_4$	5-7	100	520	60	2.1	Water	
13	$TiCl_4$	35	150	450	60	0.3	Water	
14	$TiCl_4$	5	5	400	60	0	Water	0.04μm Ti dispersed over surface

TABLE 2

GAS PHASE DEPOSITION USING FLOWING GAS AND CYLINDRICAL SYMMETRY (samples nickel foil cylinders 10 cm², water cooled cathode)

Sample No.	Gas	Pressure m bar	Current mA	Potential Volts	Time min	Weight Gain mg	Results of Electron Microprobe Analysis
15	$TiCl_4$	7-10	50	510	48	23.1	0.2μm Ti over surface with peaks up to 0.4μm
16	$TiCl_4$	4-5	50	480	44	17.6	0.1μm Ti dispersed over surface
17	$TiCl_4$	1-2	50	520	60	-2.8	No Ti detected (Much of deposit flaked off surface)
18	$TiCl_4$	8	50	450	60	3.5	0.02μm Ti over surface with peaks up to 0.06μm
19	$TiCl_4$	5-6	50	460	60	5.3	0.05μm Ti dispersed over surface
21	$TiCl_4$	7-8	50	440	60	6.3	0.02μm dispersed over surface
22	$TiCl_4$	9-10	50	480	60	8.7	0.06μm Ti dispersed over surface

a temperature of between —10 and —12°C using a mixture of crushed ice and ethanol. The difference in temperature between the reservoir and the cold trap ensured a continuous flow of vapour through the reaction zone.

For experiments involving mixtures of gases with the volatile metal halide the reaction vessel was slightly modified. The metal halide vapour was admitted into the cell through a capillary, which served to regulate the rate of flow of the vapour into the system. A side inlet was provided above the top end of the capillary below the cathode holder so that a carrier gas could be admitted and mixed with the metal halide vapour. The rate of flow of carrier gas into the system was regulated with a needle valve and measured with a differential manometer; that of the halide was determined by calibration of flow through the capillary in a given time by weight loss. Vapour pressure variations with temperature were considered when required. During operation a continuous flow of reacting mixture through the reaction vessel was thus achieved.

The vacuum system, including the halide reservoir, was flushed with argon gas before each experiment and was evacuated to less than 1 mbar pressure before admitting the vapour.

The cathode samples were weighed both before and after electrolysis in order to determine the weight of material deposited and thus obtain an estimate of the thickness. After removal from the cathode holder and weighing the cylindrical samples were unrolled and flattened in order to examine their surfaces.

A choke was placed in series with the high voltage supply to the anode to improve the stability of the circuit and reduce arcing between the electrodes.

The chemicals used were BDH GPR Grade. The surface composition was determined by X-ray fluorescence spectroscopy and microprobe analysis; surface morphologies were observed by scanning electron microscopy.

RESULTS AND DISCUSSION

1 $TiCl_4$

The results obtained with the flat coupon geometry are shown in Table 1. The weight of product deposited increased with current (samples 1, 2 and 3) but at high current densities the weight of the samples decreased with time as can be seen from the results for samples 4 to 6. Samples 4, 5 and 6 were run under identical conditions and the different weight losses observed give an indication of the percentual scatter in the results obtained in these experiments. The reason for the poor reproducibility in these preliminary experiments was probably the lack of an adequate control of the gas composition. Nonetheless, the loss in weight observed for the high current experiments (4 to 6) was due to sputtering, as a black deposit was formed on the walls of the reaction vessel which gave a positive identification test for iron. Also, the results in argon (samples 7 and 8) give a clear indication of the sputtering rate that is expected under similar glow discharge power conditions. Experiments 1 to 8 were performed without any cooling and by comparison with later experiments where a thermocouple was included in the holder the sample temperature was estimated to be in the range 400-500°C. It has been suggested that sputtering can be partially regarded as a thermal process of evaporation of the metal from regions of high temperature produced by impact of the high energy ions;[12] therefore a decrease in the temperature of the sample should alleviate this problem. That this is the case is shown in

experiments 9 to 14, from which it is clear that the control of the sample temperature is important in controlling the sputtering rate.

The results of the runs with the water cooled coupons shown in Table 1 indicate that somewhat less titanium was deposited per unit of current than in the case of the uncooled samples. It could be argued that the deposition on the uncooled samples may in part be due to thermal decomposition of $TiCl_4$ on the sample surface. In order to check this effect an experiment was devised in order to determine whether thermal decomposition of $TiCl_4$ would take place in the absence of a discharge. Firstly, a nichrome wire spiral was heated to redness by passing an electric current through it while it was in an atmosphere of $TiCl_4$ at a pressure of 6 mbar. After applying 5.1 watts to the wire for 1 hour, however, no increase in weight was observed. Secondly, a steel coupon, similar to those used in the deposition experiments, was mounted on a small electric heating element and heated twice in an atmosphere of $TiCl_4$ vapour at 6 mbar pressure. A power unit of 5.4 watts was first applied to the coupon for 1 hour, but no increase (or decrease) in its weight was observed. A power input of 22 W (enough to cause sputtering in a glow discharge) was then applied to the coupon for one hour but here again no change in its weight was observed after this treatment. These results suggest that the $TiCl_4$ is being decomposed by ion and electron bombardment in the discharge itself and that the deposition of titanium does not result from thermal decomposition of $TiCl_4$ on the surface.

The comparison of the results of experiments 1, 2 and 3 with those of experiments 13, 11 and 12 appears to indicate that the relevant variable in determining the deposition rate is the applied potential rather than the current density. This is an important conclusion; as previously mentioned, the overall applied potential will determine the electron gas temperature and, for a constant rate of energy transfer, the non-equilibrium gas temperature. In this case the applied voltage will be equivalent to the temperature for a homogeneous chemical reaction, though with the added complication that at high voltages sputtering due to ion bombardment of the surface becomes predominant. This effect can be clearly seen by comparing the results of experiment 10 with those of 11 and 12 (see later).

The results obtained with a cylindrical geometry are shown in Table 2. The effect of pressure can be seen in experiments 15 to 17, and a decrease in pressure results in a decrease in deposition rate. At low pressures, however (sample 17), it appears that sputtering becomes important and the reason for this is probably the increase in mean free path rather than any chemical kinetic effect. The consequence of this is an increase in the average energy of the positive ions bombarding the surface and hence in the sputtering rate.

The data of experiments 18 to 22 is more in accordance with a simple first reaction order gas kinetics interpretation. These results correspond to an applied potential for which the sputtering rate was small.

It can be concluded therefore that sputtering phenomena are of prime importance in determining the deposition rate for whatever reaction is occurring, and high pressures and low voltages appear to be necessary to achieve deposition. The deposits obtained were nodular (Figures 5 and 6) and XRF analysis indicated that they were strongly contaminated with chlorides (Figure 7). However, the addition of argon as a carrier gas resulted in a decrease of the Cl and an increase of the Ti XRF signals, as is shown in Figure 7. In this figure, the Ni peak corresponds to the substrate, which

TABLE 3

GAS PHASE DEPOSITION USING FLOWING GAS MIXTURES (cylindrical geometry for a water cooled cathode; samples made of nickel foil of 10 cm² area)

Sample No.	Gas	Pressure m bar	Current mA	Voltage V	Time min	Weight gain mg
27	Ar, 0.4% TiCl₄	8	25	220–260	1440	1.5
28	Ar, 1% TiCl₄	3.5	25	480	1232	49.9
30	Ar, 0.3% TiCl₄	8.5	25	300–340	1537	30.5
38	Ar, 10% H₂, 2% TiCl₄	1.5	25	580–690	1119	458.7
39	H₂, 1.5% TiCl₄	2	25	400–480	120	120.1

Sample No.	Gas	Pressure m bar	Current mA	Voltage V	Deposition Rate mg/h cm²	Average Temperature °C
23*	CrO₂Cl₂	7–8	50	450	1	400
57	H₂, 2.5% CrO₂Cl₂	4	50	480–500	3.8	120
58	H₂, 2.5% CrO₂Cl₂	4	50	490–510	3.7	120
59	H₂, 2.5% CrO₂Cl₂	4	50	490	3.8	120
54	H₂, 2.5% CrO₂Cl₂	4	50	490–500	3.9	390
55	H₂, 2% CrO₂Cl₂	5	50	490–500	3.7	390
56	H₂, 2.5% CrO₂Cl₂	4	50	500	4.0	390
69	H₂, 0.6% CrO₂Cl₂	4	75	410	1.1	350
70	A, 20% H₂, 1.2% CrO₂Cl₂	8	75	390	1.1	350

* Deposited on nickel foil; in the rest of the experiments, steel cylinders were employed.

TABLE 4

GLOW DISCHARGE CHROMIUM DEPOSITION

corroborates the non-uniform character of the deposit indicated by Figures 5 and 6.

Table 3 summarises some of the results obtained using a non-reactive (argon) and a reactive (hydrogen) carrier gas. The presence of argon results in an increase in deposition rate; for instance, the deposition rate for samples 28 and 22 are 0.24 and 0.87 mg/h cm^2, although the concentration of $TiCl_4$ in the latter case was larger by more than two orders of magnitude. This could indicate that the deposition phenomena observed are entirely due to thermal effects rather than surface neutralization and deposition of positive ions. The presence of an inert carrier gas like argon probably ensures a better energy transfer to $TiCl_4$. If the phenomenon observed is a thermal effect the effective average temperature in the discharge will be proportional to the cathode fall potential and a characteristic Arrhenius plot should be observed. To test this idea the deposition rate of Ti from $Ar-TiCl_4$ mixtures was measured at different applied potentials, while keeping the total pressure in the range 7-11 mbar; the equivalent Arrhenius plot is shown in Figure 8. From these results it is tempting to conclude that from the point of view of the reaction rate responsible for deposition the applied voltage across the cathode region is proportional to the average non-equilibrium temperature that the gas will attain. If this is correct, this result is of practical importance in enabling the determination of operating conditions favourable to a high deposition rate.

The analysis of the reaction mechanism is complicated by the several alternative reactions that can occur. If the elementary product of the thermal decomposition of $TiCl_4$ is $Ti(g)$, this can enter in various reactions before deposition occurs, i.e.,

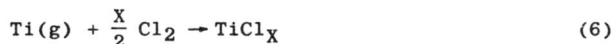

$$Ti(g) + TiCl_4 \rightarrow 2TiCl_2 \tag{5}$$

$$Ti(g) + \frac{X}{2} Cl_2 \rightarrow TiCl_X \tag{6}$$

Reaction (5) is the most likely gas phase decay mechanism of $Ti(g)$, and the value of G^o_{298} for this reaction calculated from literature data[27] was -439 kJ/mole. Therefore the deposition rate will be affected by the reactive mean free path of the species generated in the discharge.

The thermodynamics of the deposition reaction are of great importance in the glow discharge reaction kinetics as shown by the results when hydrogen was present in the reaction mixture (samples 38 and 39). The enhancement of the rate of deposition is due to the lower temperature required to drive reaction (2), as discussed previously.

2 CrO_2Cl_2

From the arguments previously presented the glow discharge deposition of chromium from CrO_2Cl_2 should be expected to occur readily. Some of the results obtained with this compound are shown in Table 4. The experiments with hydrogen as a reactive carrier gas were performed with steel cylinders instead of nickel foil. The samples were cleaned by sputtering with pure hydrogen prior to deposition; during this operation, the samples were taken to the operating deposition temperature, which was measured with a thermocouple in thermal contact with the outside surface of the cathodes.

As expected from the thermodynamic data in Figure 2 and the $TiCl_4$ results, Cr deposition was observed from CrO_2Cl_2 in the absence of a carrier or reducing gas (sample 23). The temperature of the sample had no influence on the Cr deposition rate, as can be seen by comparing the results for

samples 57, 58 and 59 with those for samples 54, 55 and 56. This result is in accordance with the previous discussion and the deposition mechanism is only related to the gas phase reactions in the glow discharge. The addition of argon to the reaction mixture did not appear to have any significant influence in this case (samples 69 and 70). The most likely explanation is that hydrogen acts as the energy transfer media between the electrons in the glow discharge and the chromium halide gas, and its partial replacement by argon does not alter significantly the coupling between the two non-equilibrium systems.

The relative ratio of hydrogen to chromyl chloride was important in determining the degree of contamination by chloride in the deposit, and essentially Cl free chromium deposits were obtained for CrO_2Cl_2 contents of 0.5% in the reaction mixture.

A nodular morphology of the deposit was often observed (Figure 9), although in some samples a columnar growth appeared. The adhesion of the deposited chromium to the steel substrate was variable; for example, Figure 10 shows some flaking of the deposit after allowing the sample to age for several weeks, and in all cases where this type of long term deterioration was observed the presence of corrosion products was apparent underneath the fractured regions. It is clear from these observations that sufficient control over the deposition variables, state of the surface, etc., was not achieved but nevertheless the reactive glow discharge deposition of chromium was certainly successful. The observation of Archer[6] regarding an increase in adhesion with increased substrate temperature was also observed in the present work.

CONCLUSIONS

The glow discharge deposition process as observed appears to be almost, if not entirely, due to chemical vapour deposition. The material which is deposited on the cathode is generated by transfer of electron energy to the metal halide vapour molecules. The chemical transformations occur presumably in the cathode fall region where the electron energy is highest and not on the cathode surface itself where, in any case, the temperature would be too low for the reactions to occur. The products of the deposition then condense onto the cathode to form the deposit. The surface temperature of the cathode influences only the adhesion of the deposit which appears to be more adherent to hot cathodes than to those which are cooled in accordance with the observations of other workers.[6]

The cathode region extends over the entire cathode surface and therefore the technique has high throwing power which is advantageous for the deposition of refractory coatings on complicated-shape substrates.

The use of volatile metal compounds as source materials is an attractive possibility for achieving deposition of a wide range of materials. Also, it seems to be possible to control the process by altering the glow discharge conditions, the pressure, the flow of gas and the composition of the gas mixture. Besides these considerations, the most important advantage of this technique is the possibility of achieving results similar to those obtained by conventional CVD but at lower substrate temperatures.

The deposition rates obtained in unoptimised experiments were about 5 μm/hour, i.e. inter-mediate between ion plating 10-1000 μm/hour[28] and sputtering 0.5-5 μm/hour.[29] The power input to the system is fairly low; values in the range of 2.5 to 4 W cm^{-2} have been employed successfully.

5 Titanium deposition from TiCl$_4$ at 4.5 mbar
   pressure; 50 mA, 480 V. Deposition time =
   44 min. Uncooled sample, cylindrical
   geometry

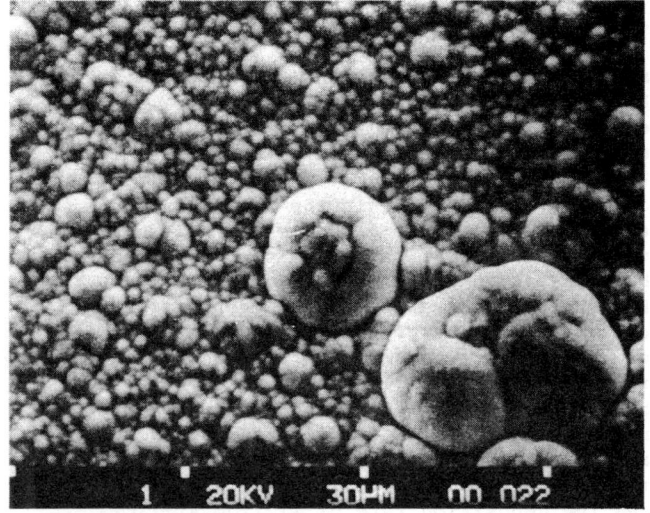

6 Titanium deposition from Ar, 0.3% TiCl$_4$ at
   8.5 mbar pressure; 25 mA, 320 V. Deposition
   time = 2½ h. Uncooled sample, cylindrical
   geometry

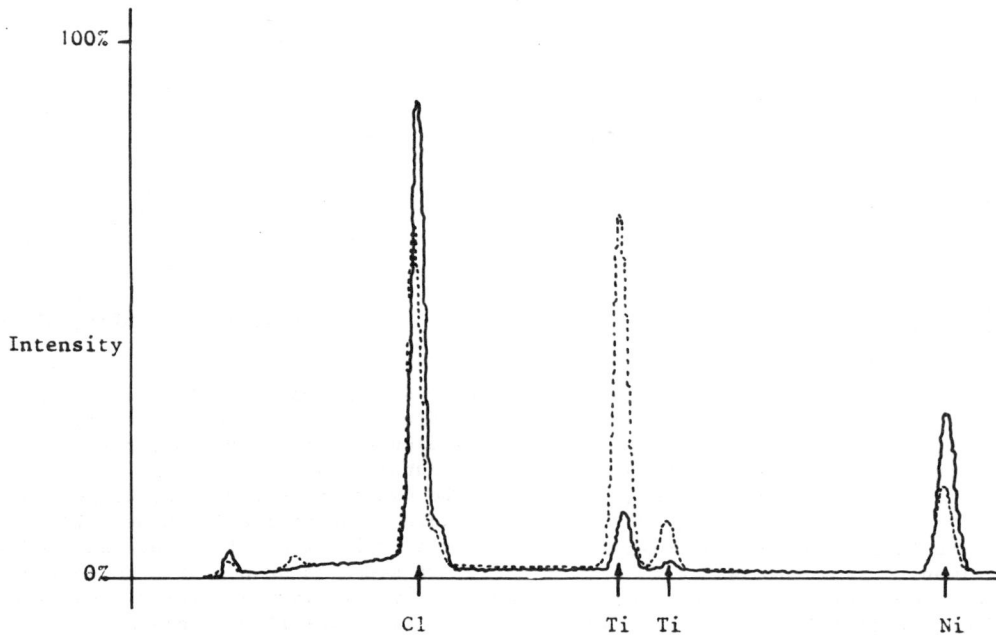

7 X-ray fluorescence spectra of titanium deposits obtained from
   ——— pure TiCl$_4$ and ----- Ar, 10% TiCl$_4$

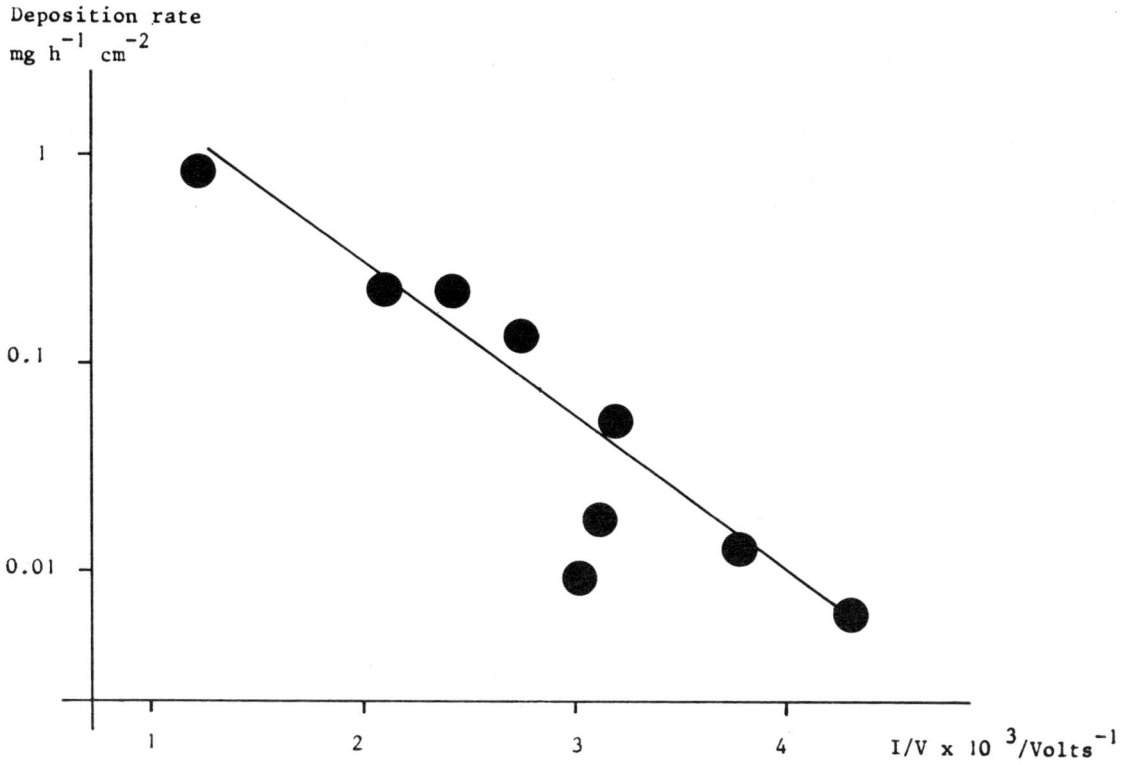

8 "Arrhenius" type plot of deposition rate as function of 1/V for deposition of Ti from mixtures containing Ar, 0.4% TiCl$_4$. Pressure 7-11 mbar.
V = total voltage applied to cell

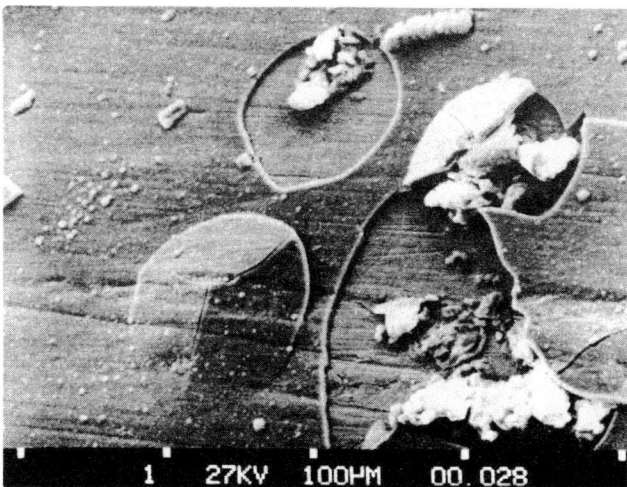

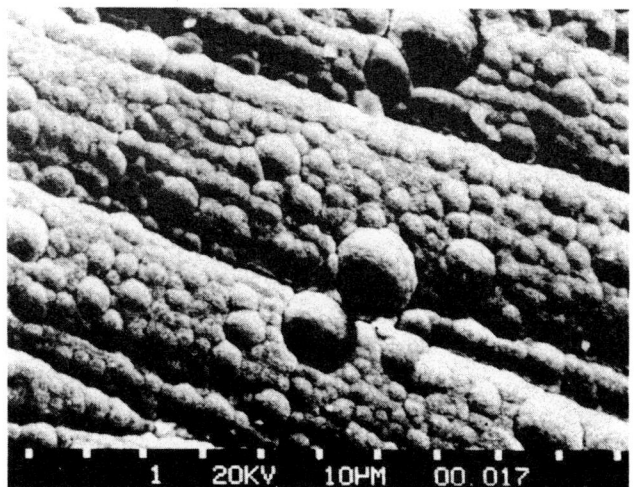

9 Chromium deposition from H$_2$, 2.5% CrO$_2$Cl$_2$ at 4 mbar pressure; 50 mA, 490-500 V. Deposition time = 1 h. Sample temperature = 390°C. Cylindrical geometry.

10 Chromium deposition from H$_2$, 2.5% CrO$_2$Cl$_2$ at 4 mbar pressure; 50 mA, 490-500 V. Deposition time = 50 min. Sample temperature = 390°C. Cylindrical geometry.

The details of the process need further study but it seems clear that irreversible reactions can be carried out in the glow discharge, even some like the thermal decomposition of $TiCl_4$ which requires temperatures in excess of 600 K.

ACKNOWLEDGMENTS

The authors are indebted to Professor G. J. Hills for proposing this area of research and for helpful discussions. The assistance in sample analysis and advice given by Dr. I. Gibson and A. Sheward of R.A.R.D.E., Sevenoaks, Kent, is gratefully acknowledged. This work has been carried out with the support of Procurement Executive, Ministry of Defence.

REFERENCES

1 Vapor Deposition, ed. C. F. Powell, J. H. Oxley and J. M. Blocher, John Wiley & Sons Inc., New York, 1966.

2 I. E. Campbell, C. F. Powell, D. H. Nowicki and B. W. Gonser, Trans. Electrochem. Soc., 96, 318 (1949).

3 J. J. Lander and L. H. Germer, Am. Inst. Mining and Met. Engrs., Inst. Metals Div., Metals Technol., 14, No. 6, Tech. Pub. No. 2259, (1947).

4 W. Hänni and H. E. Hintermann, Thin Solid Films, 40, 107 (1977).

5 I. E. Campbell and C. F. Powell, The Iron Age, 113 (1952).

6 N. J. Archer, Thin Solid Films, 80, 221 (1981).

7 H. F. Sterling and R. C. G. Swann, Solid State Electron., 8, 653 (1965).

8 J. R. Hollahan, J. Electrochem. Soc., 126, 930 (1979).

9 F. D. Egitto, J. Electrochem. Soc., 127, 1354 (1980).

10 K. Doblhofer and W. Dürr, J. Electrochem. Soc., 127, 1041 (1980).

11 A. von Engel, Ionized Gases, Clarendon Press, Oxford, 1965.

12 F. Llewellyn-Jones, The Glow Discharge, Methuen & Co. Ltd., London, 1966.

13 R. P. Stein, Phys. Rev., 89, 134 (1953).

14 T. B. Reed, Adv. High Temp. Chem., 1, 259 (1967).

15 T. B. Reed, J. Appl. Phys., 32, 821 (1961).

16 A. S. Kana'an and J. L. Margrave, Adv. Inorg. Chem. Radiochem., 6, 143 (1964).

17 A. G. Massey, J. Chem. Educ., 40, 311 (1963).

18 T. Wartik, R. Moore and H. I. Schlesinger, J. Amer. Chem. Soc., 71, 3265 (1949).

19 D. K. Lam, R. F. Baddour and A. F. Stancell, J. Macromol. Sci. Chem., A10, 421 (1976).

20 C. Boelhouwer and H. I. Waterman, Research (London), Supplement 9, S11 (1956).

21 W. W. Cooper, H. S. Mickley and R. F. Baddour, Ind. Eng. Chem., Fundamentals, 7, 400 (1968).

22 J. C. Viguié in Science and Technology of Surface Coating, ed. B. N. Chapman and J. C. Anderson, Academic Press, London, 1974, p. 145.

23 C. W. Dannatt and H. J. T. Ellingham, Dis. Faraday Soc., 4, 126 (1948).

24 N. Sano and G. R. Belton, Metall. Trans., 5, 2151 (1974).

25 F. A. Miller, G. L. Carlson and W. B. White, Spectrochim. Acta, 15, 709 (1959).

26 JANAF, Thermochemical Tables, Nat. Stand. Ref. Data Ser., Nat. Bur. Stand. (U.S.), 2nd Ed., 1971.

27 D. Altman, M. Farber and D. M. Mason, J. Chem. Phys., 25, 531 (1956).

28 P. A. Higham and D. G. Teer, Thin Solid Films, 58, 121 (1979).

29 R. A. Dugdale, Thin Solid Films, 45, 541 (1977).

CLAUDIA WIECZOREK

Plasma enhanced chemical vapour deposition of tantalum on silicon

SYNOPSIS

With the plasma enhanced chemical vapour deposition (PECVD) process thin tantalum films were deposited on silicon and sapphire substrates at temperatures of about 450°C. Under these conditions the tantalum nucleates in the high-resistivity β-phase (170 $\mu\Omega$cm). After heat treatment (900°C, 1 h) in argon atmosphere we could identify α-Ta on sapphire, and $TaSi_2$ on silicon substrates by X-ray diffraction. In both cases a considerable decrease in resistivity could be observed.

THE AUTHOR

The author is with
Siemens Research Laboratories in Munich.

INTRODUCTION

In the semiconductor technology the application of a CVD process is commonly used for depositing silicon and passivation layers (SiO_2, Si_3N_4). The advantages of this method are the high purity and the good step coverage of the films. Substrate temperatures up to about 1000°C are necessary. However, if the substrate is thermo-sensitive or even if the previous existing layers give rise to reaction or diffusion a lower deposition temperature has to be chosen. One possible method to realize this is the plasma enhanced chemical vapour deposition (PECVD).

As gate contact material, e.g. for MOS-FET-transistors highly doped polycrystalline silicon is used. For VLSI (very large scale integration) application the resistivity of this material is too high ($\varrho \approx 750$ $\mu\Omega$ cm). Therefore it has been proposed to overcoat the thin poly-Si layer with a low-resistivity silicide film (e.g. $TaSi_2$, $MoSi_2$, WSi_2) by sputter deposition or evaporation.[1] The total thickness of this sandwich layer is 500 to 800 nm.

Silicide also can be produced by a solid state reaction between silicon and a refractory metal (e.g. tungsten) above 600°C.[2] In this case both materials can be deposited by a CVD process.

Since tungsten easily forms volatile oxides which may cause contamination of the semiconductor material we studied the deposition of tantalum. This metal forms a stable oxide passivation layer (Ta_2O_5) with a low partial pressure even at high temperatures.

APPARATUS

In constructing an apparatus for CVD one must remember that tantalum has a great affinity to oxygen. For this reason care had to be taken to avoid oxygen impurities. So the whole apparatus, including the gas lines, was constructed for high vacuum. The leakage rate was less than 0.1 μbar l/s.

From a central gas supply we receive the carrier gas hydrogen and argon for ventilating the reactor. The water content and the oxygen content in H_2 (Ar) is less than 1 (2) ppm respectively.

Fig. 1 shows a schematic diagram of our reduced-pressure CVD system which is well suited for process development.[3] As starting material we used tantalum pentachloride (99.98 % $TaCl_5$). The $TaCl_5$ is filled in a vacuum tight stainless steel sublimator under nitrogen atmosphere. To establish a definite $TaCl_5$ partial pressure all pipes and valves between the sublimator and the reactor were heated to a slightly higher temperature than the sublimator. Since corrosive by-products are generated during the reaction, all metal pieces were made of stainless steel.

A plasma of the reaction gas was excited by an induction coil which was supplied by an rf-generator (3 to 5 MHz) with a maximum power of 5 kW. The quartz tube reactor contained an electrically isolated substrate holder for a 3 " silicon wafer, so its potential was floating. The substrate could be warmed up by resistance and/or induction heating. The substrate temperature was measured with a NiCr-Ni thermo-couple.

With our pumping system (Roots pump and rotary pump with a mechanical and chemical oil filter) a residual gas pressure of 1 μbar could be established. Since the pumping capacity could be regulated by a throttle valve a total pressure up to 10 mbar could be established even with small amounts of carrier gas.

EXPERIMENTAL

By depositing CVD tantalum on silicon at temperatures of about 850°C, tantalum silicide ($TaSi_2$) with a rough surface is formed. Since films with such a high surface roughness cannot be used in semiconductor technology we tried to produce a smooth material at a lower temperature. Therefore

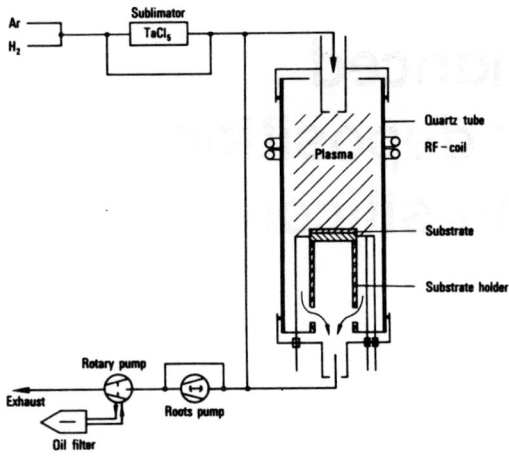

Fig. 1   Schematic diagram of the CVD apparatus

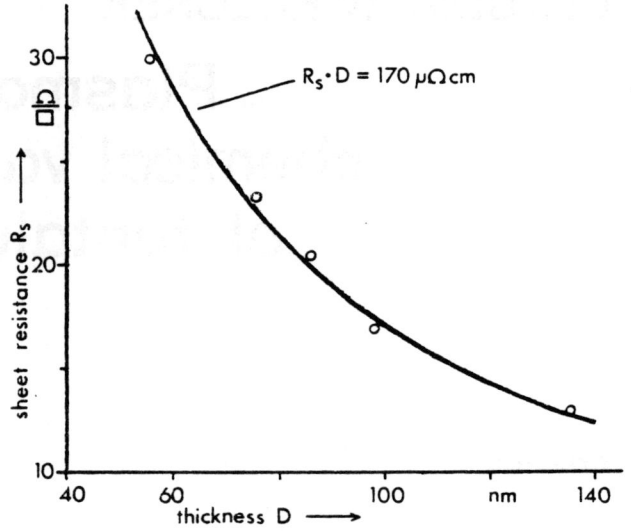

Fig. 3   Sheet resistance as a function of the β-Ta film thickness

$R_s \cdot D = 170 \, \mu\Omega \, cm$

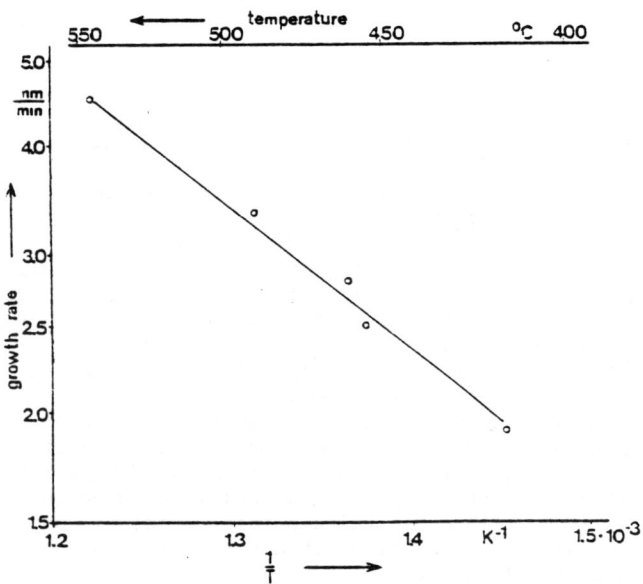

Fig. 2   Arrhenius plot for the deposition of β-Ta

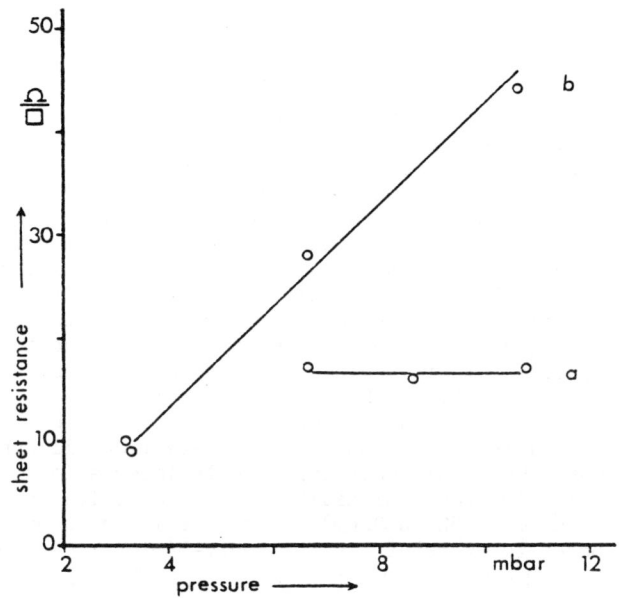

Fig. 4   Sheet resistance as a function of the total pressure during deposition at 460°C, a) high $TaCl_5$ concentration, b) low $TaCl_5$ concentration

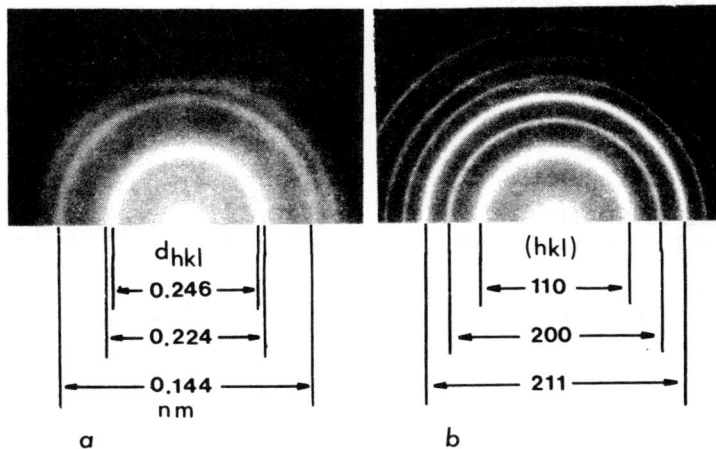

Fig. 5 Electron diffraction pattern of PECVD
Ta films; a) β-phase after deposition,
b) α-phase after annealing in Ar

we studied a plasma enhanced CVD process. The gross
formula is the same as in the case of high-temperature CVD [4]

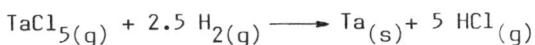

$$TaCl_{5(g)} + 2.5\ H_{2(g)} \longrightarrow Ta_{(s)} + 5\ HCl_{(g)}$$

Typically 0.8 l/min hydrogen flowed over the
heated $TaCl_5$ (T≤120°C) to carry the $TaCl_5$ vapour into the plasma zone. With the throttle
valve the pressure can be varied between 2 and
10 mbar, a typical value was about 5 mbar.

The substrates were (100) silicon wafers.
For some studies sapphire (α-Al_2O_3) substrates
or apertures coated with a very thin amorphous
silicon-oxide film for electron-optical investigations were used.

In studying the annealing behaviour of the
PECVD tantalum, some samples were heated up to
900°C in a pure atmosphere of argon, nitrogen
or forming gas (N_2/H_2).

The thickness of the tantalum films on silicon and on sapphire measured by a stylus method[+]
was up to 200 nm; on electron microscopical
apertures the film thickness was about 30 nm.
The sheet resistance was determined with a con-
tactless eddy current method.[++]

The structure of the films was identified by
electron and X-ray diffraction.

RESULTS

Our experiments were performed with an input
power of the rf-generator of about 2 kW and a
frequency of 3.5 MHz.

Highly reflecting tantalum films were obtained at substrate temperatures between 400
and 500°C. A maximum growth rate of 10 nm/min
could be reached. From the data of Fig. 2 an
activation energy of 30 kJ/mol was estimated for
the deposition of tantalum. The films were identified as β-Ta by electron and X-ray diffraction.
From the thickness dependence of the sheet resistance (resistivity/thickness) a resistivity
of about 170 $\mu\Omega$cm can be assumed (Fig. 3).
The uniformity of the sheet resistance measured
at nine points across the 3 " wafers was better
than ± 10 % for all samples.

At a constant substrate temperature (460°C)
the total pressure was varied for two ratios of
$TaCl_5/H_2$. As one can see from Fig. 4a at a high
concentration of $TaCl_5$ the reaction was nearly
independent of the total pressure within the
range investigated. However, if the concentration of $TaCl_5$ was lower, the growth rate decreased corresponding an increase in sheet
resistance with increasing pressure (Fig. 4b).

On annealing a Si/Ta sandwich layer at 900°C
in argon atmosphere, the tantalum reacts to
$TaSi_2$. The resistivity of this silicide is
lower than 100 $\mu\Omega$cm. On silica substrates
or sapphire the β-Ta transforms to α-Ta during
this heat treatment (see electron diffraction
patterns of Fig. 5).

[+] Tencor   Alpha-step 2
[++] Tencor   Sonogage rt², m-gage

Heating up the β-Ta in a nitrogen or in a forming gas atmosphere the reaction of β-Ta with nitrogen forming TaN (cubic) is faster than the reaction of tantalum with silicon. If the annealing atmosphere is not free of oxygen, the formation of Ta_2O_5 is favoured compared with the formation of TaN, $TaSi_2$ or α-Ta.

DISCUSSION

With the PECVD process highly reflecting tantalum films can be deposited on silicon at about 450°C. The relatively low resistivity of about 170 $\mu\Omega\,cm$ in comparison to UHV evaporated β-Ta films [5] may be attributed to the higher deposition temperature of the PECVD process. The reason why tantalum nucleates in the β-phase is not yet fully understood. Perhaps incorporated impurities like H_2, Cl_2 or Si stabilize this phase up to 800°C.[6] Whether it is possible to produce α-Ta films - analogous to sputtered films - by using a substrate bias voltage [7] must be studied in further experiments.

ACKNOWLEDGMENT

The author wishes to thank Dr. O. Eberspächer and Mrs. J. Reischl for the X-ray analyses.

REFERENCES

1   D. B. Fraser, S. P. Murarka, A. R. Tretola, A. K. Sinha, J. Vac. Sci. Technol., 18 (1981) 2, 345-348

2   N. Hashimoto, Trans. Met. Soc. of AIME, Vol 239 (1967), 1109-1111

3   M. Stolz, K. Hieber, C. Wieczorek, Thin Solid Films, to be published

4   K. Hieber, Thin Solid Films, 24 (1974), 157-164

5   A. Schäfer, G. Menzel, Electrocomponent Sci. Technol., Vol 4 (1977), 29-35

6   W. D. Westwood, N. Waterhouse, P. S. Wilcox, Tantalum Thin Films, Academic Press 1975

7   K. Hieber, E. Laute nbacher, Thin Solid Films, 66 (1980), 191-196

M BOOTH, M LEES, and A M STAINES
Development of a plasma carburising process

SYNOPSIS

A comparison of a glow discharge carburising process with conventional carburising techniques has been undertaken. An experimental glow discharge apparatus has been developed and is described. Carbon concentration profiles have been obtained using the glow discharge treatment and are compared with similar data for vacuum and gas carburising. The results show the glow discharge process compares most favourably with the alternative methods. A number of benefits associated with this new process are presented.

THE AUTHORS

Dr Booth & Mr Lees are research officers at the Electricity Council Research Centre, Capenhurst. Mr Staines is Manager of the Wolfson Plasma Processing Unit at Birmingham University.

INTRODUCTION

Thermochemical treatments of steels are widely used for improving the wear and fatigue properties of components. In particular, nitriding and carburising have become accepted methods of surface hardening. New developments have continued to be made in these processes, the main objectives being (i) to improve the metallurgical properties of the component, (ii) to reduce the overall process costs and (iii) to reduce the environmental impact of the waste products from the process.

Among the advances made in the last decade has been the commercial development of a sub-atmospheric pressure nitriding process termed "ion nitriding" or "plasma nitriding". This treatment uses an anomalous glow discharge and operates at pressures upto 12 mbar with nitrogen hydrogen mixtures or cracked ammonia as the treatment gas.
A number of advantages over conventional treatments accrue from glow discharge processing. These include a metallurgically superior product, in addition to savings in energy and treatment gas[1-5]. Clearly, the development of a glow discharge carburising process with these inherent advantages would be of great commercial significance. This is emphasised when the size of the respective markets is considered. Despite a number of investigations over the last few years[5-10], no significant progress has been reported. Research is currently being undertaken at the Electricity Council Research Centre (ECRC) and at the Wolfson Plasma Processing Unit (WPPU), Birmingham University which is aimed at assessing the characteristics of glow discharge carburising and specifying the requirements for a commercial system.

2. CHARACTERISTICS OF THE GLOW DISCHARGE

The carburising treatment is carried out in a direct current, anomalous glow discharge created in a methane/hydrogen gas mixture at a pressure in the range 1-25mbar. Current densities of $50-100A/m^2$ are used and the discharge sustains a voltage of 600-700 volts at the processing temperature. The workpiece is connected as the cathode in the discharge system, whilst the chamber walls constitute the anode and are at earth potential.

Although the properties of glow discharges have been the subject of a number of texts, see for exampl von Engel[11] and Llewellyn-Jones[12], it is instructive to recall the voltage distribution within the discharge system, figure 1. It is seen that the majority of the potential drop occurs close to the cathode, that is, within the cathode fall region. In this region a range of particle interactions can occur under the influence of the high electric field from which ions and, via collisions, molecules acquire high kinetic energy. Thus the number of physical phenomena occurring in the discharge are many and varied, some of which are listed below.

(i) Sputtering – the bombardment of the workpiece surface by molecules and ions with high kinetic energy, which produces a clean surface and reduces the effect of passive layers on reaction kinetics at the gas/solid interface.

(ii) Heating effects – a result of translational kinetic energy of particles and molecules impinging on the workpiece surface.

(iii) Mass transfer effects – metastable chemical species may be produced in the highly active cathode fall region and particle flux to the workpiece surface is accelerated by virtue of the high electric field.

(iv) Uniformity of treatment – for surfaces with a characteristic dimension which is large with respect to the cathode dark space, the electric field, and hence the acceleration of ions, is

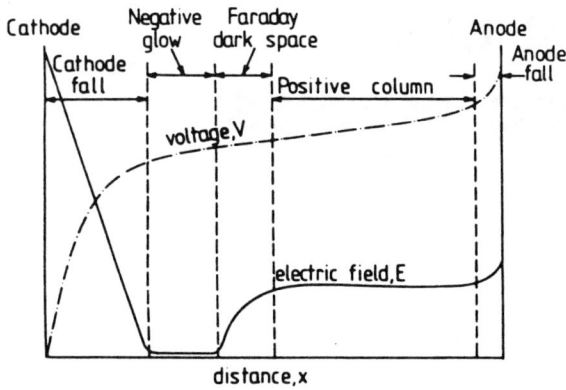

1. Schematic diagram of spatial distribution of
potential and electric field in a glow discharge.

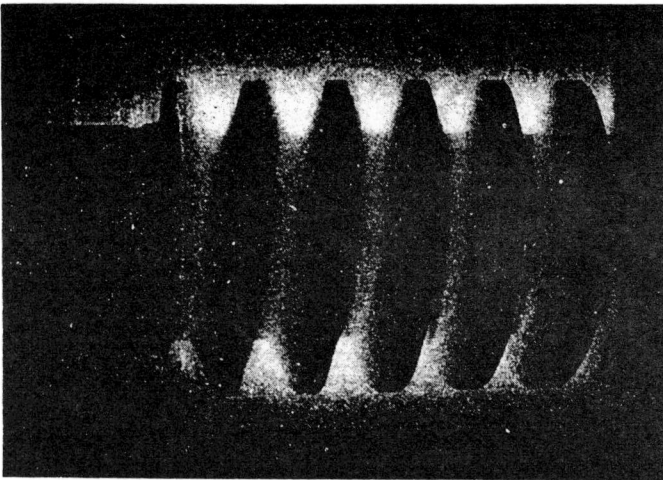

2. Samples undergoing a glow discharge treatment.

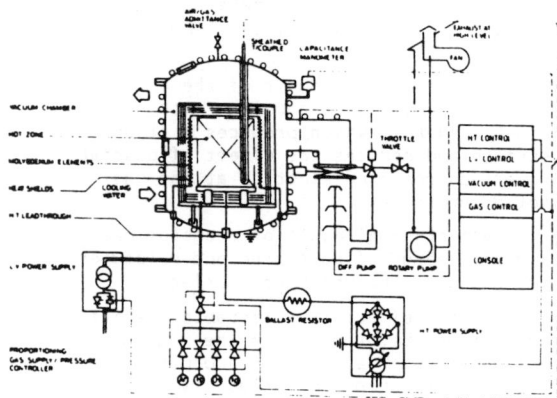

3. Schematic representation of the glow discharge
carburising system.

effectively normal to the surface.
This permits a uniform treatment to
be achieved even on complex geometries,
provided no hollow cathode effects
occur.

Figure 2 shows a component during a glow discharge
treatment and demonstrates the uniformity
discussed above. The luminous nature of the
discharge from which the term "glow discharge"
is derived, is due to excitation caused by
electrons colliding with gas molecules after
having been accelerated across the high field,
dark space region of the discharge (see figure 1).

3. ADVANTAGES OF GLOW DISCHARGE CARBURISING

A small number of investigations of glow discharge
carburising have been reported during the past
five years[5-10]. The results of these studies
indicate that the process offers several
advantages when compared with conventional gas
carburising, as summarised below.

(i) Reduced process time - substantial savings
have been observed[7-10], especially
when compared with vacuum carburising.
Greater reductions in process times have
been reported[7] when compared with gas
carburising.

(ii) Reduced energy consumption - mainly due to
reduced process time, although the use of
a cold-wall vacuum furnace with lower heat
loss through the gas exhaust also increases
the system efficiency.

(iii) Reduced gas consumption - due to the low
pressure used (in the range 1-25mbar).
Less than 1 furnace volume per hour
provides good results[6].

(iv) Improved case uniformity - a direct result
of the discharge characteristics (see
section 2), noted in all the reports[5-10].

(v) Reduced sooting - attributed to low gas
consumption and dissociation of the carburising gas occurring mainly in the cathode
fall region of the glow discharge.

(vi) High processing temperatures - the use of
hydrocarbons provides a much higher carbon
availability over carbon monoxide, especially
at high temperatures, so that glow discharge
carburising with methane allows higher
temperatures to be used to advantage.

(vii) No gas generator equipment needed - simple gas
mixtures are used in the glow discharge
process.

(viii) Environmental aspects - very few pollution
problems are experienced with the vacuum-
based glow discharge method.

4. EXPERIMENTAL EQUIPMENT

4.1 General system description

The experimental apparatus, shown schematically in
figure 3, has been designed and built at ECRC. It
comprises a standard 70 litre water-cooled vacuum
chamber, pumped by means of a rotary vacuum pump. An
oil diffusion pump has also been fitted to allow high
vacuum heat treatments to be carried out when required.
A conventional high tension leadthrough is mounted in
the base of the chamber. Radiative heat losses are
reduced by means of a molybdenum/stainless steel heat
shield with additional refractory insulation surrounding the heat shield. Resistance heated radiation

elements are arranged around the inside of the inner
heat shield, within the hot zone. The system is shown
in figure 4.

In normal operation, the workpieces are placed on a
baseplate connected to the high tension leadthrough
and heated to the treatment temperature by the combined
effect of the glow discharge and radiation heating.

Parallel work has been carried out using a 20kW glow
discharge nitriding unit at WPPU, Birmingham
University. This unit, shown in figure 5, has been
modified to enable carburising treatments to be
achieved, as described in the work of Staines et al
(15,16).

4.2 The auxiliary heating system

The benefit of using an auxiliary heating system is
the reduction of power required to be supplied by the
discharge. This means that the discharge can operate
further from the glow-to-arc transition (for example
at point A in figure 6) than in a carburising unit
heated by the discharge alone (for example at point B
in figure 6). The probability of a catastrophic
breakdown is thus reduced.

The auxiliary heating system decouples the operating
voltage and current levels of the glow discharge from
the heating requirements of the workpiece and a greater
degree of flexibility of glow discharge power input can
be achieved. Substantial improvements in temperature
uniformity are also obtained using the radiant heating
system in conjunction with the glow discharge.

An indication of the effect of the auxiliary heating
system can be obtained by comparing the equilibrium
temperatures of two samples placed in similar positions
within the hot zone, with one electrically isolated from
the cathode. For a typical heat treatment, a simple
calculation shows the auxiliary heating system supplies
twice the heating power of the glow discharge.

4.3 Control Equipment

4.3.1 Pressure/Gas flow measurement and control

Because of the need to vary gas composition and control
the pressure (see Section 5) a means of sensing the
pressure which is independent of the gas composition
is required. This rules out the use of a pirani
gauge. Instead, a capacitance manometer was used and
the pressure was dynamically controlled by varying
the flow of the main constituent gas of the treatment
atmosphere. The control equipment allows the mixing
of up to four gases. Currently these are hydrogen,
methane, argon and nitrogen. The flow of any of these
gases can be set independently of any other gas, or
alternatively can be a set fraction of the flow of
the main constituent gas.

The vacuum system is fully controlled by means of a
simple 3-way switch which is linked to all the
relevant pneumatically operated valves and pressure
sensing heads.

4.3.2 Temperature Measurement and control

Temperature measurements are obtained at present by
means of chromel/alumel thermocouples placed in
alumina sheaths which are embedded in the test pieces.
However, difficulties using this method of temperature
measurement in the glow discharge process have already
been recognised (17,18) and the possibility of using
a non-contact temperature measurement system is being
considered.

Control is achieved by use of a standard temperature
controller. The output of the controller drives the
supply of the resistance heating elements.

5. PROCESS DETAILS

A typical process sequence is shown in figure 7. The
sample is heated to the process temperature in a non-
carburising atmosphere with the auxiliary heating
system operating at maximum power. During this period,
the pressure is gradually increased to the set value
for the carburising process and the discharge power
level is simultaneously increased in order to minimise
the heat up period. As the operating temperature is
approached, the discharge power is reduced to a
pre-determined level and the system is allowed to
stabilise at temperature before commencing the
carburising treatment.

The start of carburising is marked by the introduction
of the carburising atmosphere, generally a methane/
hydrogen mixture at a preselected composition. At
the end of the treatment time, the glow discharge
and heating elements are switched off and the samples
are allowed to cool at pressures up to 130 mbar using
a range of gases.

6. COMPARISON WITH OTHER CARBURISING PROCESSES

Figure 8 compares carbon concentration profiles
obtained by vacuum carburising at various pressures.
The results at the two higher pressures were
obtained by Eysell[13] and demonstrate the pressure
dependence of the carburising process. For an
atmosphere of pure methane, a pressure of at least
130 mbar is required for significant mass transport
of carbon into the steel surface. Increasing the
pressure causes an increase in the surface carbon
concentration indicating an enhanced rate of mass
transport of carbon into the steel.

Vacuum carburising at pressures similar to those
required for the glow discharge process produces
very little carburisation, as shown in figure 8.
This result was obtained using a treatment atmosphere
of 20% CH_4 /80% H_2 at 25 mbar, where the partial
pressure of methane is only 5 mbar, (substantially
lower than in the 130 mbar treatment).

The beneficial effect of using the glow discharge
for carburising is shown in figure 9. A
comparison with the carbon profiles in figure 8
shows that the glow discharge profile in figure 9
is almost identical to that obtained after
vacuum carburising at both a higher temperature
(1040°C) and higher methane partial pressure (260 m
bar). Indeed, there is almost a three orders of
magnitude decrease in the methane partial pressure
used in the glow discharge treatment.

Figure 10 compares conventional gas carburising data
(14) with carbon concentration profiles obtained by
the glow discharge process. This data emphasises
the increased process rate at directly comparable
carburising temperatures obtained when using the
glow discharge. Preliminary results[15,16]
indicate that this can be attributed to an increased
carbon mass transfer rate in the active, dark space
region of the glow discharge. It can be seen that
high surface carbon concentrations are rapidly
established using glow discharge carburising. An
increased carbon gradient is therefore obtained,
and a more rapid diffusion process will result.
Increasing the carburising temperature to 1000°C
using the glow discharge technique results in a five-
fold decrease in process time over that shown by the
gas carburising data. The higher operating
temperature provides a faster diffusion process
as well as an increased rate of decomposition of
the carburising gas, both of which contribute to
a more rapid carburising process.

4. The experimental glow discharge surface treatment unit.

5. A 20kW glow discharge surface treatment unit.

23.4

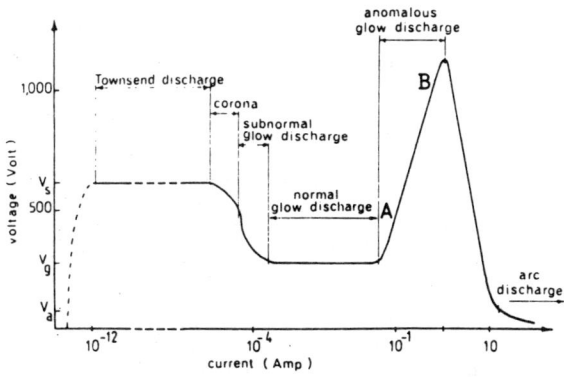

6. Typical voltage-current characteristic of a glow discharge.

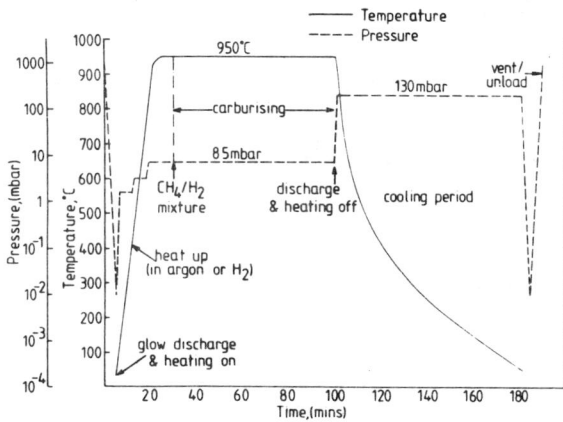

7. Schematic representation of a typical glow discharge carburising cycle.

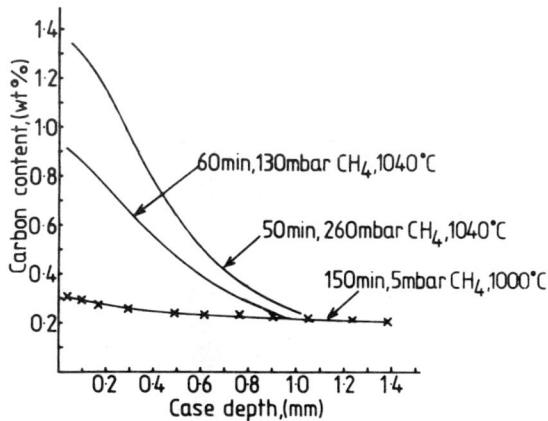

8. Carbon concentration profiles of vacuum carburised[13] low alloy steel showing the influence of methane partial pressure.

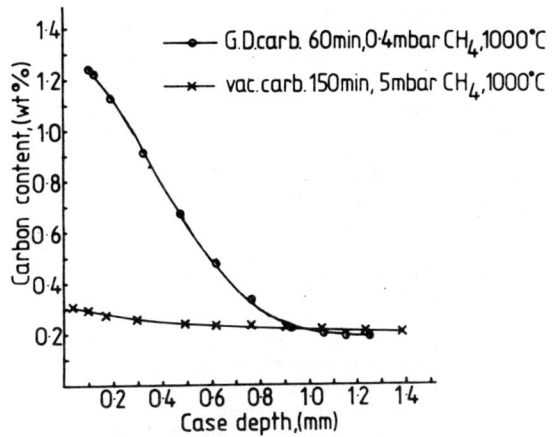

9. Comparison of carbon concentration profiles obtained by glow discharge carburising and low pressure vacuum carburising.

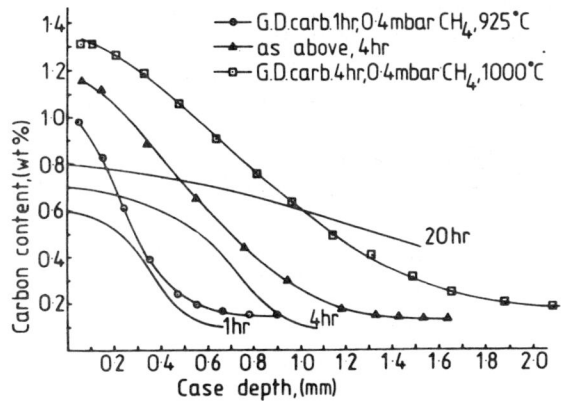

10. Carbon concentration profiles of conventional gas carburised[14] and glow discharge carburised mild steel. (Conditions for gas carburising, 0.25% CO_2 925°C, 1-20hr).

Previous workers[5-10] claim that sooting within the chamber does not occur during glow discharge carburising, in contrast to the problems which can be found in gas carburising[14]. Our own experience indicates that this is not the case. Sooting should be expected since many surfaces in the carburising chamber are at temperatures at which pyrolysis of methane can occur. The results we have obtained show that the hot zone and leadthrough systems continually acquire a sooty deposit, but this does not cause any operational difficulties at the low levels of methane used during the process.

Sooty deposits have only been observed on a limited number of workpieces, in circumstances where the rate of carbon mass transfer to the sample surface has exceeded the corresponding rate of diffusion into the sample[15,16].

7. INDUSTRIAL REQUIREMENTS OF THE GLOW DISCHARGE CARBURISING PROCESS

Optimum surface properties with conventionally carburised components are generally achieved at a maximum surface carbon level of 0.8 wt%. The results obtained with the glow discharge technique indicate that this carbon level is almost always exceeded, even at short treatment times. This indicates that the treatment gas may be too rich in carbon, even at 5% CH_4. However, the use of the 5% CH_4/H_2 gas mixture has been shown to produce a high carbon concentration profile very rapidly. A means of redistributing the carbon added in this "boost-diffuse" glow discharge carburising cycle can therefore be envisaged, in a manner analogous to that already used for gas carburising[14], but with substantial process time savings.

The process lends itself to easy pressure and gas composition control, both of which have been shown to be important parameters in determining the carbon concentration profile. Precise control of the carburising process may therefore be possible simply by adjusting the system pressure and gas composition, thus eliminating the need for sophisticated gas analysis equipment in the control system.

Commercial carburising practice requires a full hardening cycle, which means that a quench facility is needed in the system. The use of standard oil quench baths in vacuum furnaces is already well established[20], but the possibility of using a gas quench should not be overlooked.

A production glow discharge carburising system must be capable of satisfying the following criteria:

  (i)   rapid loading/unloading
  (ii) rapid pump down
 (iii) fast heat up
  (iv) low heat loss
   (v) quench facility for full hardenability
  (vi) rapid and controlled carburising

The results and experience described in this paper, together with current vacuum furnace technology demonstrate that the prospects of a glow discharge carburising system meeting the above criteria are extremely good.

8. CONCLUSIONS

1. Glow discharge carburising offers significant advantages over conventional gas or vacuum carburising.

2. A given carbon concentration profile is achieved more rapidly with the glow discharge treatment than with conventional carburising processes.

3. A greater mass transfer rate in the glow discharge process results in a higher surface carbon concentration than is achieved using conventional carburising processes.

4. A reduction in carburising gas consumption (by more than two orders of magnitude), is obtained with the glow discharge process when compared with vacuum carburising.

5. Auxiliary heating provides greater flexibility and improved temperature control in the glow discharge process

6. Sooting of chamber fittings may occur, but this does not affect the system operation.

REFERENCES

1 Edenhofer, B, "Physical and metallurgical aspects of Ionitriding", Heat Treatment of Metals (1974), 1, 23–28 and 59–67

2 Korotchenko, V and Bell, T, "Application of plasma nitriding in UK Manufacturing Industries, 1978", Heat Treatment of Metals (1978), 5, 88–94

3 Staines, A M and Bell, T, "Plasma nitriding of high alloy steels", Proc. Conf. "Heat Treatment – Methods and Media", Inst. Met., London (1979), pp 58–69

4 Marciniak, A & Karpinski, T, "Comparative studies on energy consumption in installations for ion and gas nitriding", Industrial Heating, April (1980), 42–44

5 Staines, A M & Bell, T, "Technological importance of plasma induced nitrided and carburised layers on steel", Thin Solid Films (1981), 86, 201–211

6 Grube, W L & Gay, J G, "High rate carburising in a glow discharge methane plasma", Met. Trans. A, (1978), 9A, 1421–1429

7 Grube, W L, "Progress in plasma carburising", J. Heat Treating, (1980), 1, 40–49

8 Collignon, P, Hisler, G, Michel, H, and Gantois, M, "Study of carburising and carbonitriding by ion bombardment: comparison with conventional processes", Heat Treatment '76, Metals Society, London (1977), 65–70

9 Yoneda, Y & Takami, S, "Ion carburising process for industrial applications", Proc. Conf. "Vacuum Metallurgy", Pittsburgh (Science Press, Princeton, 1977), 135–156

10 Rembges, W, "Ion carburising – vacuum carburising in a glow discharge" presented at 18th Harterei Technische Fachtagung, 11–12 Nov. 1981, Suhl, GDR

11 von Engel, A, "Ionised Gases", (Oxford University Press 1965)

12 Llewellyn-Jones, F, "The Glow Discharge", (Methuen & Co Ltd, 1966)

13 Eysell, F W, "Carburising in the low pressure range: process parameters and application", Electrowarme Int. (1976), 34, B312–317

14 Dawes, C, & Tranter, D F, "Production gas carburising control", Heat Treatment of Metals (1974), 1, 121–130

15 Staines, A M, Booth, M, & Lees, M I, "Austenitic thermochemical treatments in a glow discharge", to be presented at BNCE Conference 'Electroheat for Metals', Cambridge, 21–23 September 1982

16 Staines, A M, PhD Thesis, to be submitted, Liverpool University 1982

17 Staines, A M, contribution to the discussion, Heat Treatment '81, to be published 1982 by The Metals Society, London

18 Dixon, G J, Plumb, S, & Child, H C, "Processing aspects of plasma nitriding", Heat Treatment '81, to be published 1982 by The Metals Soc., London

19 "Carburising and carbonitriding", ASM (1977), pp 52–53

20 Ruffle, T W & Byrnes, Jnr, E R, "Quenching in vacuum furnaces", Heat Treatment of Metals, (1979), 6, 81–87

A R NYAIESH and L HOLLAND

The growth of amorphous and graphitic carbon layers under ion bombardment in an r.f. plasma

SYNOPSIS

Methods of preparing C-films from energetic C-ions and from ion impact bombardment of hydrocarbon species are reviewed. It is shown that amorphous C-deposits resulting from impact damage can have dielectric properties and be exceptionally hard. The r.f. plasma technique used to grow a-C-films on a negatively biased target is described and the effects of temperature rise on structural change followed using differential scanning calorimetry. The as grown films contain absorbed hydrogen which produces compressive stress and is released to form gas bubbles at 400°C.

THE AUTHORS

Mr. Nyaiesh is in the School of Engineering and Applied Sciences and Prof. Holland is in the School of Mathematical and Physical Sciences, University of Sussex, Sussex, England.

INTRODUCTION

Carbon films of high electrical resistivity ($\sim 10^{12} \Omega$cm) and exceptional hardness have been prepared by several techniques which involve film growth under conditions of ion bombardment which results in radiation damage of the carbon structure providing the deposit temperature does not rise sufficiently for graphitization. Such films have been referred to as diamond-like because of their hardness but structural studies in our laboratory indicate that they are amorphous.

Shown in Fig. 1 are the chief techniques developed for the deposition of hard a C-films. These are as follows:

1   Ion source deposition using a low energy C+-ion beam[2]

2   Sputter-deposition of carbon whilst exposed to bombardment from an Ar+-ion beam[3]

3   Plasma deposition using a d.c. glow discharge in a hydrocarbon gas to deposit carbon on the ion bombarded cathode[4] in the drawing shown a grid in front of the cathode provides secondary electrons which travel to the cathode to prevent the bombarded C-film from developing a negative potential and

4   plasma deposition using an r.f. glow discharge in a hydrocarbon gas to deposit carbon on a negatively biased probe which is capacitively coupled to the r.f. supply.[5]

Hard C-films have been used for: anti-reflection $\lambda/4$ coatings on Ge-optics to enhance the infra-red transmittance[6,7] optical filters on polycarbonate components[8] electron strippers in heavy ion accelerators[9,10] and they are of considerable interest as coatings on the inner walls of plasma fusion vessels and as hard coatings on metals providing also a low coefficient of friction ($\mu \approx 0.2$ for hard C-films on steel in contact). As r.f. plasma a C-films are free from pinholes and inert when in contact with most acids and solvents they could have chemical uses.[1]

The latter technique is the one considered in this paper and it has the advantage that the film substrate does not have to be a conductor and positive surface charging of the high resistivity

TABLE I

Method:	1	2	3	4	5	6	7
Initial Temp:	100.0	200.0	200.0	350.0	350.0	500.0	500.0
Program Rate:	20.0	0.0	20.0	0.0	20.0	0.0	20.0
Final Temp:	200.0	200.0	350.0	350.0	500.0	500.0	610.0
Iso Minutes:	0.0	5.0	0.0	5.0	0.0	5.0	0.0
Program Cool:	No	No	No	No	No	No	No
Program Gas:	1	1	1	1	1	1	1
Next Method:	2	3	4	5	6	7	0

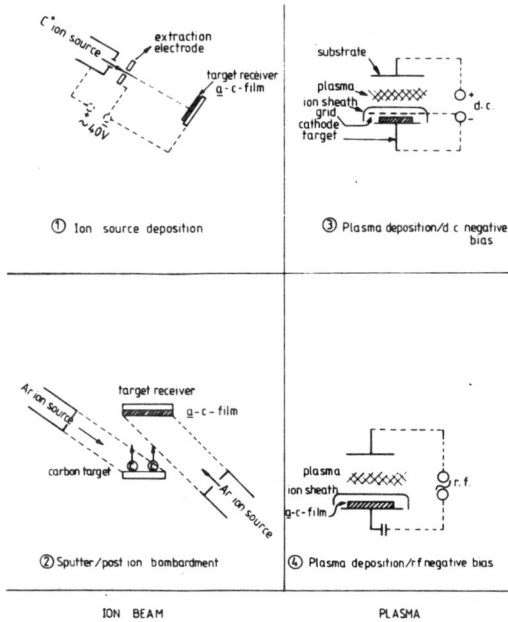

1   Methods used to prepare hard C-films

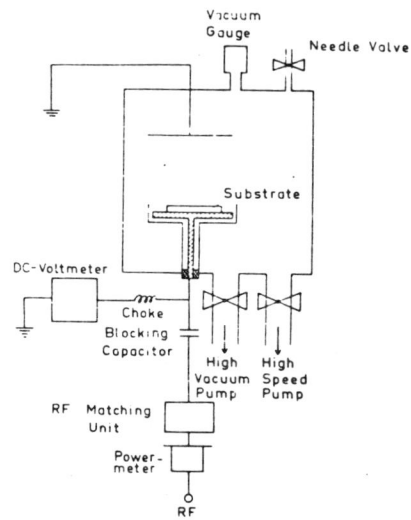

3   R.F. system for preparing hard a-C-films in a hydrocarbon plasma.

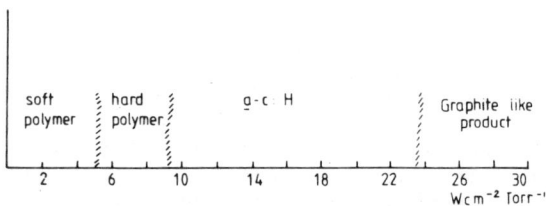

2   Dependence of film growth on r.f. power density to hydrocarbon gas pressure (CH_4).

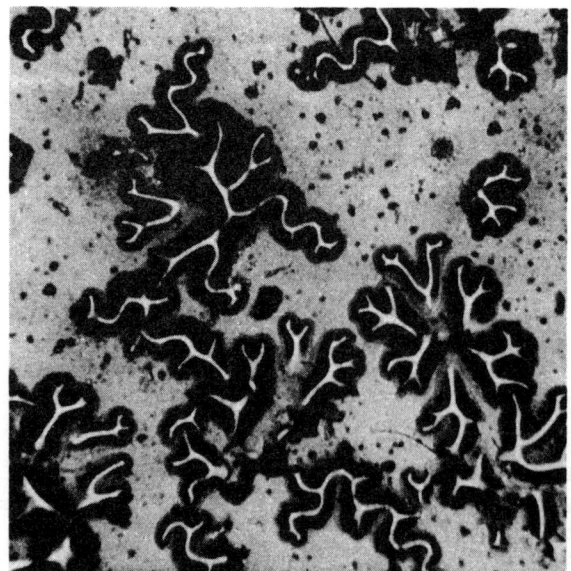

4   Wrinkling of a-C-film from relief of compressive stress after detachment from soda lime glass base.

C-film is avoided by the flow of electrons from the plasma during part of the positive r.f. cycle. Further the positive ions bombarding the film and transporting material are accelerated in a positive ion sheath which with plane and slight curved receivers can be made to give uniform current density and thus film growth over large areas.

When an r.f. glow discharge is excited in a hydrocarbon gas at low gas pressure using a capacitively coupled electrode as a receiver support both ionized and neutral components of the fragmented hydrocarbon bombard the receiver. At low power densities and rough vacuum pressures the degree of hydrocarbon dissociation both in the gas phase and from ion impact at the receiver is insufficient to rupture all C-H bonds. Accordingly a polymer film is formed from the unsaturated hydrocarbon components. As the r.f. power input is raised or the pressure lowered to an optimum value the carbon content of the film rises and although hydrogen is present in the film it appears that this is mainly buried in the free state. Evidence for this is the absence of absorption bands for C-H bonds in the infra-red transmission spectra.[6,11] If the r.f. power input is excessive then the heat energy released can result in the film graphitizing during growth. The temperature rise during growth must not exceed ∿175°C to obtain a hard C-film with an amorphous structure, whereas films subsequently heated in air can be raised to about 350°C when their hydrogen content is precipitated as bubbles.

The foregoing growth effects are shown in Fig. 2 as a function of the r.f. power input (W) per unit area of receiver (cm^2) and unit gas pressure (torr) for methane. Although the C-film structure shows some dependence on ion energy,[12] i.e. the ion sheath potential, the amorphous hard carbon form could generally be obtained at between 10 to 20Wcm^{-2} torr^{-1} for the r.f. frequencies used from 3 to 13.56MHz. However the hydrogen absorbed in these a C-films, which can be as much as 25 at.% produces a compressive stress that increases with film thickness. This stress can result in a weakly adherent film being detached from its base. It appears that base materials known to form stable compounds with carbon, e.g. Si, Ti, etc. are adherent to a C-films.

The temperature stability of a C-films is obviously an important consideration when determining their possible uses and this is the subject of this paper.

It was considered that differential scanning calorimetry (DSC) was a simple technique for following structural changes arising from temperature rise and an account is given of the DSC characteristics of a C-films containing hydrogen.

EXPERIMENTAL

The apparatus used to prepare the a C-films is shown in Fig. 3. The r.f. power supply was connected to a water-cooled electrode via an L-matching network and a blocking capacitor. The remaining r.f. electrode and the vessel base and top plate were all grounded. The difference between the forward and reflected r.f. power, i.e. the power input, was measured with a Bird wattmeter (Model 43). The reflected power was kept to a negligible level by adjustment of the L-network. The negative d.c. bias to ground developed at the receiver electrode was measured

with a voltmeter with a series choke to provide a high r.f. impedance.

The receiver to be coated, in the experiments, a microscope glass slide, was rested on the r.f. electrode. Hydrocarbon gas (CH4) was admitted to the vessel via a needle valve. The deposition conditions used in the experiments were : methane pressure = 10^{-1}torr, power input allowing for 1W reflected power = 31W and negative bias potential to ground = 420V.

Carbon films grown in the above manner became wrinkled and detached from the glass base as shown in Fig. 4 when a few thousand angstroms thick. The deposit was collected from several runs and placed in a pan (0.5cm dia.) and covered with a lid sealed by a press. The sample capsule, made of aluminium, was then placed in the Ar-oven of a DSC instrument (Du Pont Instruments thermal analyser model 1090) with a similar but empty capsule in the reference position.

RESULTS AND DISCUSSION

The DSC was set to raise the sample and reference capsules at two constant rates of temperature rise, 10 and 20°C min^{-1}, over the temperature region 80 to 600°C. Under these conditions no difference in the rate of supplying heat energy to the capsules was observed up to 200°C as shown in Fig. 5 for a rise rate of 10°C min^{-1} for a 0.3mg sample of a C.

The common capsule temperature and the heat energy rate difference between the reference and sample capsules were automatically recorded on a floppy disc and this information was used to integrate the total difference in the heat flowing to the capsules.

A second a C sample of 0.21mg was sealed in a capsule and the DSC curve obtained at 20°C min^{-1} temperature rise. The results given in Fig. 6 show that a structural change is occurring with a peak rate at about 370°C. A second run with the same sample also gave a peak showing that the structural change was incomplete. The latter coincides with the temperature at which the resistivity of plasma a C-films falls and absorbed hydrogen is precipitated as bubbles in the solid. It is believed that a C-films have a metastable structure so that heat treatment will cause a transition to the lower energy graphitic form. At normal temperature hydrogen cannot diffuse in the a C-structure but as the temperature is raised structural changes begin to allow hydrogen movement and gas bubbles appear at the DSC maximum heat flow temperature. To reconfirm this an a C-film was heated in an Ar oven and gas bubbles were observed at ∿400°C.

It was possible that structural changes at low temperature were time dependent but were not observed because of the rate of temperature rise. Thus an a C sample of 2.39mg was heated at 20°C min^{-1} but as the temperature reached given values it was held constant for 5min intervals commencing at 200°C; the heating programme is given in Table 1.

The DSC curve for the temperature region 150 to 600°C given in Fig. 7 shows that there is no discontinuity in the heat flow characteristic but the temperature for maximum heat flow has risen to 590°C. It is believed that the periods of prolonged heating at a constant temperature

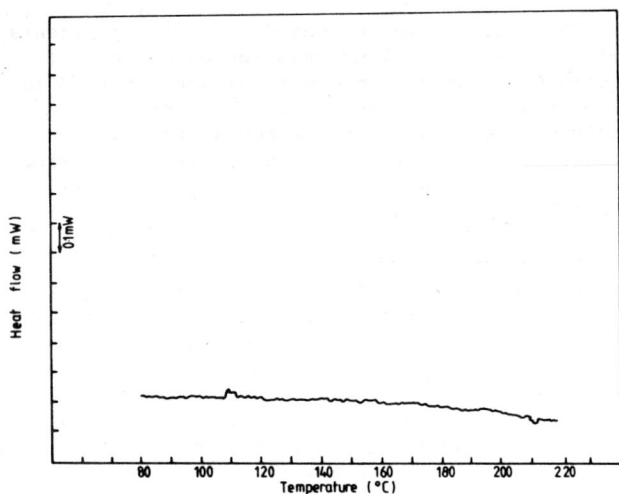

5 DSC curve for 0.3mg C 10°C/min 80 to 220°C.

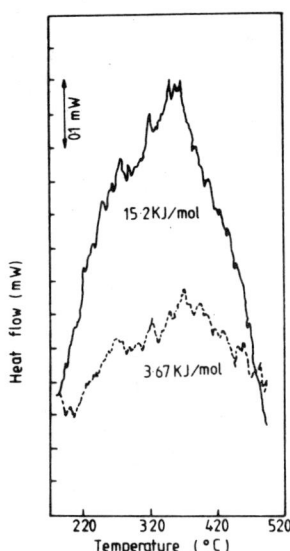

6 DSC curve for 0.21mg C 20°C/min 200-520°C;
the lower curve is the second run.

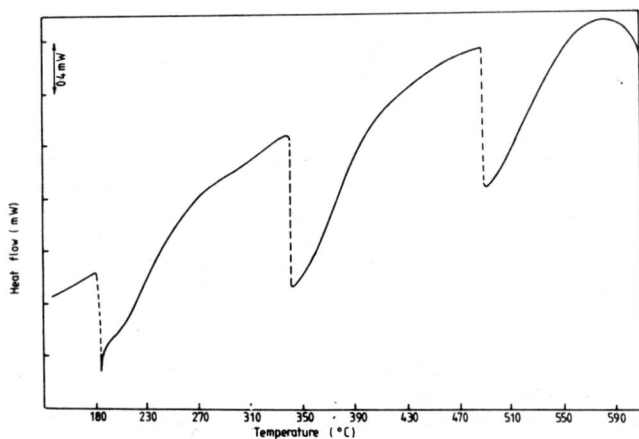

7 DSC curve for 2.39mg C 20°C/min 150-600°C
with const. temp. intervals of 5min.

resulted in greater film annealing than with a continuous temperature rise but as some structural changes have a high activation energy this has the effect of raising the temperature maximum.

Integrating the area under the heat flow curve for the DSC sample in Fig. 6 and using a molecular weight of 12 (assuming only C) gives an activation energy of 15.2KJ mol^{-1}.

R.F. plasma film grown under the conditions described here have been examined by Berg and Andersson[13] using X-ray photo-emission spectroscopy and they found that the emission spectrum for the insulating hard C-films had features which placed them between diamond and graphite. Holland and Ojha[11] from their study of physical properties concluded that this intermediate state tended to be amorphous from ion impact damage, i.e. providing energy released during growth did not cause annealing. Thus it will be apparent from this and previous work that plasma grown a C-films containing hydrogen can be used in applications where their unusual physical and chemical properties are of value providing they are not likely to be heated above about 200°C because of their metastable structure.

REFERENCES

1 L. Holland and S. M. Ojha, Thin Solid Films, 38 (1976) L17-19

2 S. Aisenberg and R. Chabot, J.Appl.Phys., 42 (1971) 2953-2958

3 Chr. Weissmantel, Thin Solid Films, 58 (1979) 101-105

4 D. S. Whitmell and R. Williamson, Thin Solid Films, 35 (1976) 255-261

5 L. Holland, U.K. Pat. 1582231 (Appl. Aug.1976)

6 L. Holland and S. M. Ojha, Thin Solid Films, 48 (1978) L21-23

7 A. Bubenzer, B. Discher and A. R. Nya⁺⁻ Thin Solid Films to be publis⁻

8 A. Nyaiesh and L. Holland, "Amorphous Carbon Films on Polycarbonate Substrates", 4th Internat. Colloquium on Plasmas and Sputtering, Nice, Sept. 1982, Soc. Franc. Du Vide

9 N. R. S. Tait et al., Nuclear Instrum. Meth., 163 (1979) 1

10 N. R. S. Tait, D. Tolfree, P. John, J. M. Odeh and M. Thomas, Nuclear Instrum. Meth., 176 (1980) 433

11 L. Holland and S. M. Ojha, Thin Solid Films 58 (1979) 107-116

12 S. M. Ojha, H. Norström and D. J. McCulloch, Thin Solid Films, 60 (1979) 213-225

13 S. Berg and L. P. Andersson, Thin Solid Films, 58 (1979) 117-120